MANUFACTURING RESEARCH AND TECHNOLOGY 10

MANUFACTURING PLANNING AND CONTROL

MANUFACTURING RESEARCH AND TECHNOLOGY

MANUFACTURING RESEARCH AND TECHNOLOGY 10

Manufacturing Planning and Control

A Reference Model

by

Frank P. M. Biemans
North American Philips Corp., Briarcliff Manor, NY, U.S.A.

ELSEVIER
Amsterdam – Oxford – New York – Tokyo   1990

ELSEVIER SCIENCE PUBLISHERS B.V.
Sara Burgerhartstraat 25
P.O. Box 211, 1000 AE Amsterdam, The Netherlands

Distributors for the United States and Canada:

ELSEVIER SCIENCE PUBLISHING COMPANY INC.
665 Avenue of the Americas
New York, NY 10010, U.S.A.

ISBN 0-444-88647-8 (Vol. 10)
ISBN 0-444-42505-5 (Series)

Printed in The Netherlands

TP

Preface

Modern production organisations are required to efficiently and quickly produce a wide variety of products with increasingly shorter life cycles. Managing such organisations is a complicated task.

A primary reason for the complexity is the lack of clarity as to how modifications of *components* of the production organisation contribute to the performance of the organisation as a *whole*. In other words, what ultimately matters is the 'bottum line' efficiency and flexibility of the overall production organisation. However, people within the production organisation, such as product designers, material managers, maintenance personnel, and process engineers, deal with components of the overall organisation. How do changes of the components affect the organisation as a whole?

This book focuses on the above question and outlines solutions based on concepts from Information Science and Systems Theory, knowledge of manufacturing as an application domain, and experience with the design of computerised factory control systems. More specifically, it describes the development of a 'Reference Model', which is a representation of an idealised production organisation, defining the global tasks of its components as well as the relations between the components and the whole.

The contribution of this work therefore lies in the construction of a systems view of a production organisation, encompassing production planning, inventory management, scheduling, machine control, servo control, sensing, product design, process planning, machine design, configuration management, and more.

As an illustration, this work also includes pointers to complete functional specifications for components in a computerised production organisation. It should be kept in mind, however, that the Reference Model has also found application as a reference for the analysis and rationalisation of non-automated or partially automated production organisations.

This work differs from the mainstream literature on manufacturing. Unlike most publications from the "Computer Integrated Manufacturing" community, it does not emphasise the use of computers. Rather, it focuses on and provides solutions for the essential planning and control problems, irrespective of the means that can be applied to execute planning or control tasks.

Similarly, this work differs from much of the reported work on Operations Research, which largely describes algorithms for isolated planning and control problems; problems that sometimes do not even exist. Instead, this work identifies the boundary conditions, objectives, and information requirements for such algorithms so that they are relevant in the context of the production organisation as a whole.

The development of the Reference Model starts with a description of a production organisation as a black box, from the perspective of a commercial department. This commercial department outlines the production targets and budgets for one or more production organisations and is therefore interested in their overall efficiency and flexibility improvements. Subsequently, in a well defined sequence of steps, components of the black box production organisation are identified so that these components can jointly realise the required black box behaviour. Thus, it becomes clear how components of the production organisation contribute to the task and performance of the production organisation as a whole.

This work addresses people studying the control organisation and procedures in factories, people responsible for the overall design of computerised manufacturing control systems, and developers of algorithms for manufacturing planning and control. To a lesser extent, this book also addresses designers of large-scale systems in general.

Contents

II Reference Model for MPCSs   53

4 Decomposition of an MPCS   55

5 Decomposition of a Cell/Line   83

Acknowledgements

First and foremost, I would like to acknowledge my Ph.D. advisor, Chris Vissers, University of Twente, The Netherlands. His scholarship, critical questioning, scrupulous proofreading, insistence on justification of designs, and directions have been crucial for developing the Reference Model presented in this book.

I also owe deep thanks to Sjoerd Sjoerdsma, Philips CAM Centre, for the many fruitful discussions and advises, but also for his perseverance and friendship in pursuing a cooperative effort to introduce and improve MPCSs.

This book is based on my work in industrial laboratories, rich in people and viewpoints, under constant pressures, and with varying degrees of success. The many viewpoints were a source of inspiration, the pressures an incentive to improve, and our mistakes lessons for the future.

In particular, I wish to thank about 25 colleagues at Philips Laboratories–Briarcliff, Philips CAM Centre, Philips I&E Industrial Automation, and several Philips Product Divisions, who have built factory control systems on the basis of the Reference Model, who have promoted the Reference Model, or who have compared existing control systems with the idealised one described by the Reference Model. The applications and real-life problems have been the major motivation to pursue a better theoretical understanding and widely applicable solutions. I also wish to thank the managers that were involved for their interest, support, and criticism: Ron Benton, Roger Goffin, Jo van den Hanenberg, Ernie Kent, Loek Nijman, Mark Rochkind, and Cor Scholten.

Pieter Blonk (University of Twente) and Krishnakumar Bhaskaran (Philips Laboratories–Briarcliff) are acknowledged. They spent many hours discussing Reference Models, as a result of which many ideas matured, and also did a significant amount of proofreading.

Many thanks to the people who were so kind and patient to share their knowledge of factories: the foreman, schedulers, manufacturing engineers, process engineers, product designers, equipment engineers, operators, plant managers of Airpax in Cambridge, MD, and Cheshire, CT (USA); Computer Peripherals International in Norristown, PA (USA); Canon, NKK steel, Suntory, Sharp IC, Sony, and Matsushita Electric Company in Japan; Océ, Grolsch, Nijverdal ten Cate, Volvo Car, Willem II, and Vredestein in The Netherlands; Hewlett Packard in West-Germany; Digital Equipment Corporation in Malboro, MA (USA); DOW Chemical, in Midland, MI (USA); Ford in Detroit, MI, and in Landsdale, PA (USA); Philips Lighting in Bath, NY (USA), Aachen (West-Germany), and Weert (The Netherlands); Kulka Wiring in Mount Vernon, NY, (USA); Philips Consumer Electronics in Krefeld, Wezlar (West Germany), Eindhoven (The Netherlands), Brugge, Hasselt (Belgium), and Greeneville, TN (USA); Philips Small Domestic Appliances in Drachten, Hoogeveen, and Groningen

(The Netherlands); Philips Components in Tilburg, Eindhoven, Nijmegen (The Netherlands), and Dreux (France); Philips Major Domestic Appliances in Amiens (France); Philips Machine Factories in Eindhoven, Sittard (The Netherlands), and South Plainfield, NJ (USA), Philips Circuit Assemblies in Milwaukee, WI (USA); and Signetics in Sunnyvale, CA (USA).

I also wish to express my gratitude for the support of Anne Marie Biemans. Her cheerfulness and enthusiasm ensured that this work, which can be quite addictive, would take up only a fraction of our live. Her understanding and encouragement made it possible that this work, which can also be frustrating at times, could continue. Finally, I wish to thank my parents for their care, support, and encouragement.

Briarcliff Manor, NY
Frank P.M. Biemans

Part I

Introduction

Chapter 1

A Systematic and Methodical Approach to Manufacturing Control

1.1 Manufacturing Efficiency and Flexibility

Modern factories are required to respond efficiently to production targets, which call for varying numbers and types of products with increasingly shorter life cycles. We will discuss systems that can control these factories, the 'Manufacturing Planning and Control Systems' or 'MPCSs'. They accept production targets and manufacture products accordingly, provided that they are supplied with the resources needed to manufacture these products.

MPCSs, with various degrees of automation, have been in existence since the industrial revolution. However, they could be improved, could they not? Note that companies pay dearly for their huge inventory, their inability to quickly translate product concepts into manufacturing reality, their inability to meet customer demand for diversified products, the inadequate utilisation of their machines, the failure to automate some of their work, etc.

We may improve the manufacturing efficiency and flexibility by eliminating non-value adding activities, simplifying product designs, monitoring critical process parameters, redesigning the layout, doing preventive maintenance, etc. These are necessary activities to rationalise and streamline the manufacturing processes. However, they can help us only to a limited extent because they cannot address the rationalisation of the *control* system, which, after all, has the supervision of the manufacturing processes.

Integrated Manufacturing Control. We have the expertise to control a transport system, or a storage system. But, we tend to view such systems in isolation. Consequently, the MPCS components, which we develop to control these systems, are stand alone components, unable to contribute to the efficiency and flexibility of the entire MPCS, which consists of many components.

We may improve, for example, an MPCS component that controls a storage system so that it can respond more quickly to requests to dispatch products. However, this improvement may have no effect on the MPCS as a whole if we do not change, for example, the MPCS components that could exploit the improved responsiveness to the dispatch requests, such as the components that issue them.

To be able to make large scale performance improvements to factories, we need to know how to build an integrated MPCS.

An 'integrated MPCS' is an MPCS that consists of components that efficiently contribute to the task, functional behaviour, and performance of an MPCS as a whole.

To build an integrated MPCS, we should know which decisions an MPCS has to make, and how these decisions interrelate and affect the overall MPCS. However, as Figure 1.1 illustrates, we know very little about the interrelationships of these decisions and the performance and flexibility of the MPCS as a whole. This makes it seemingly impossible to realise the goal of an integrated MPCS.

How to organise all the, human and automated, MPCS components in a factory so that the performance and flexibility of the entire MPCS is improved? How is the control of inventory levels related to the allocation of jobs to workstations, or to the movement of robot arms, or to the movement of parts in a transport system, or to the sensing of a temperature, or to the recognition of an object? What are the relevant control tasks and how do they relate to each other? What need to be considered when making a control decision, and what can be ignored? Which algorithm or procedure is preferable? How quickly should a control decision be made, and how good should it be? Who is responsible for maintenance? Engineering or production? What are useful performance indicators for a foreman so that he contribute to the "bottum line" performance of the overall MPCS?

We have no direct answers to these questions. We therefore deal with 'islands of control' rather than with the control of a factory as a whole. Hence, we cannot assign tasks to humans or machines in a fashion that will efficiently contribute to the overall MPCS. That helps explain why one can see situations such as a commercial department and a factory that both forecast the production targets without proper coordination; poor throughput because scheduling possibilities are not exploited; logistic systems without accurate data about the shop floor; product quality flaws that are detected long after they have been caused; or products that are well designed but cannot be manufactured. It also explains why one can find people and systems spending more

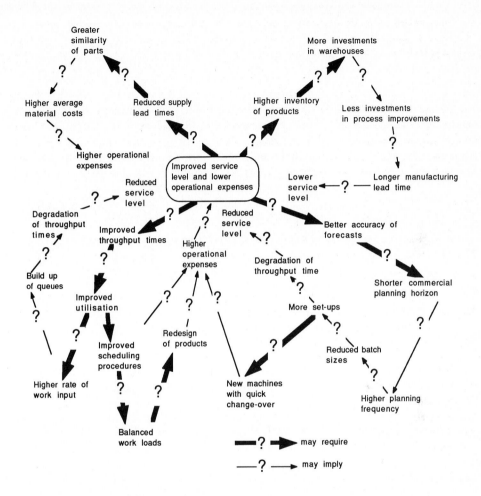

Greater
similarity
of parts

More investments
in warehouses

Higher average
material costs

Reduced supply
lead times

Higher inventory
of products

Less investments
in process improvements

Higher operational
expenses

Improved service
level and lower
operational expenses

Reduced
service
level

Lower
service
level

Longer manufacturing
lead time

Degradation
of throughput
times

Improved
throughput times

Reduced
service
level

Better accuracy of
forecasts

Higher
operational
expenses

Build up
of queues

Degradation of
throughput time

Shorter commercial
planning horizon

Improved
utilisation

More set-ups

Improved
scheduling
procedures

Redesign
of products

Reduced batch
sizes

Higher rate of
work input

New machines
with quick
change-over

Higher planning
frequency

Balanced
work loads

━━?━➤ may require

──?──➤ may imply

Manufacturing Control: many parameters can be changed but the overwhelming number of causal and non-causal relations between these parameters, which are not well understood in terms of their magnitude and timing, makes it difficult to determine which parameters should be changed to achieve overall performance improvements.

This figure presents some parameters that may have to be changed to improve the service level of an MPCS, that is how well it can deliver products on time, while reducing its operational expenses. It shows several loops of potential cause and effect relations. Due to these loops, most changes intended to have a positive effect on the service level and operational expenses, ultimately may have negative effects as well.

Figure 1.1: A Labyrinth Caused By A Myriad Relations

time on coordination than on doing actual work, doing what could be done better by others or is already being done by others, or improving individual performance in an environment that cannot catch up.

It should be emphasised that the problems sketched above are largely caused by our negligence or inability to validate the relevance of planning and control tasks in the context of the overall MPCS, or, conversely, because of our negligence or inability to analyse which tasks are necessary to improve the performance and flexibility of the entire MPCS. We invest heavily in systems that do not really fit in their environment because they create redundant, inefficient, or sub optimising 'islands of automation'. We lack an integrated perspective. Note that 'integration', as used here, should not be confused with integration at a physical level by means of computer networks or computer busses. Rather, we address the semantics of integration –the information that MPCS components should share.

The above observations and the benefits of large scale performance improvements to factories motivate us to search for techniques to build an integrated MPCS. In addition, we search for techniques that allow us to build MPCSs that are flexible.

A 'flexible MPCS' is an MPCS that, in addition to realising production targets, can quickly and efficiently honour requests to change its product portfolio, production capacity, or production costs.

While we do not quantify our definition of a flexible MPCS, we look for techniques to build MPCSs that are significantly more flexible than today's MPCSs. Today's MPCSs often consist of dedicated components, or of flexible components that are so intricately interwoven with each other that the overall MPCS is still inflexible.

In the following Sections, we will roughly outline some major requirements regarding the approach to build integrated and flexible MPCSs.

1.2 Systematic Approach

We propose a systematic approach to design MPCSs: define the required function of an MPCS as a whole, identify its components and determine the required relation between the whole and the components, and subsequently determine how to implement these components. This approach will reveal a systems perspective of manufacturing control.

A 'systems perspective of manufacturing control' reveals the relation of an MPCS as a whole and its components in terms of their tasks, functional behaviour, and performance.

This systems perspective is of immense practical relevance because it helps to determine how changes in MPCS components affect the overall efficiency and flexibility of an MPCS.

The literature tends to abstain from taking a systematic approach and therefore fails to provide solutions that are guaranteed to improve the overall performance of an MPCS.

> **Example.** The upper part of Figure 1.2 illustrates a scheduling paradigm typically reported in literature. The problem is to realise n jobs by m machines so as to minimise, say, lateness. All jobs are ready to be executed and all machines are idle, the jobs are broken down in operations to be scheduled, the schedule is computed, the operations are executed by the machines, and all machines are idle again. The literature focuses on the combinatoric optimisation to find a satisfactory sequence of operations and machine allocation; the relevance of the problem definition in the context of an MPCS typically is not validated.
>
> The bottum part of Figure 1.2 illustrates the systems perspective on scheduling that we will develop in the course of this work. We will discuss an MPCS component, whether human or automated, responsible for scheduling. It has to execute algorithms that enable it to reply to requests for the execution of jobs in certain time slots, quickly assess whether it can make commitments regarding the execution of these jobs, break jobs into operations and schedule these operations, accommodate requests to execute jobs at any moment, provide status reports about the progress of jobs, re-reschedule to accommodate for the stochastic behavior of machines, etc.
>
> This systems perspective reveals some requirements for the algorithms to be executed by the scheduling components: start dates of jobs should be taken into consideration, "on-the-fly" arrivals of jobs should be allowed, and feed back from the machines should be taken into consideration. Such requirements are not considered in the typical scheduling paradigms. If one further studies the scheduling context, MPCS components will appear that generate scheduling algorithms on the basis of performance feed back regarding the actual execution of the algorithms.

Like theoreticians, many practitioners fall short in taking a systematic approach. They suggest, for example, that computer networks be used to connect controllers with the intention of providing flexibility because controllers can communicate with each other. They do not define the information to be passed via these networks. Without a systems view, it is not clear which information should be exchanged by the controllers.

Rather than taking a systematic approach, one tends to approach manufacturing as an application domain for specific techniques like linear programming, scheduling, bar

8

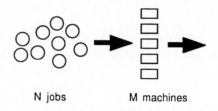

N jobs   M machines

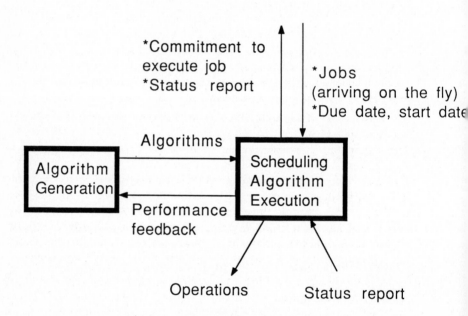

*Commitment to
execute job
*Status report

*Jobs
(arriving on the fly)
*Due date, start date

Algorithms

Algorithm
Generation

Scheduling
Algorithm
Execution

Performance
feedback

Operations   Status report

Figure 1.2: An Isolated Versus A Systems Perspective On Scheduling

code reading, or data communication – a tendency to seek a problem that fits a solution. One relies too heavily on the applicability of a specific technique or solution to be effective in a domain as complex as manufacturing. This indicates a need for research: while the technology for manufacturing is improving rapidly, a basic understanding of the systems issues remains incomplete [50].

1.3 Methodical Approach

We would also like to have a methodical approach to design an MPCS, a well-founded, structured, consistent, and procedural approach.

Today's manufacturing is largely ad hoc. Intuition plays a more predominant role than rational procedures and often experience gathered in one factory is put to little use in another. In short, a text book "how to control a factory ?" does not exist. As a result, it takes a long time to develop systems and procedures to control a specific factory. Moreover, it is not clear whether the procedures and systems in a specific factory can be improved.

This book provides a methodical approach to design integrated and flexible MPCSs.

We will assume that we can design an MPCS from scratch; there is no existing MPCS that can only be modified to a certain extent so as to avoid disturbing the on-going production. Our MPCS is an 'idealised' MPCS, intended as a normative description of an MPCS. We will therefore take a seemingly unrealistic approach and design an idealised MPCS from scratch rather than discussing existing MPCSs. Such a design, while unlikely to be fully implemented, can serve as a reference for the modification of an existing MPCS, as a target for its evolution. It can further serve as a basis for its evaluation.

1.4 Uniqueness

Why is it that there is a lack of integrated perspective in MPCSs? There are several possible explanations.

Lack of Resources. Manufacturing is yet to be fully regarded as a subject for scientific study and analysis. Consequently, there are not many researchers in the field of manufacturing science and still fewer who have a comprehensive understanding of various facets of manufacturing. This does not contribute to the development of an integrated perspective of manufacturing control.

Discipline-Oriented Approaches. As suggested earlier, most scientists, practitioners, and text books in manufacturing take a discipline-oriented approach.

The typical discipline-oriented approaches rely too heavily on the applicability of a specific technique or solution to be effective in a domain as complex and multidisciplinary as manufacturing. They certainly do not lead to an integrated perspective of control.

A systematic approach, which first attempts to define the required function of components of an MPCS and subsequently looks at several technical disciplines to implement the required functionality, has better chances to lead to applicable results.

Design Strategy. The formidable complexity of manufacturing has another implication as well. Other authors have already argued that the design of a complex system can best be done by a single mind to achieve the logical consistency that is required from a system in order to be used efficiently [26,33]. This requirement seems still not sufficiently understood in the technical and scientific communities. This may help explain why many large projects with similar goals, typically funded by US or European governments, have not led to precise, complete, and unambiguous proposals for flexible and integrated MPCSs.

Introduction Strategy. What could explain the absence of real-life, integrated MPCSs? One is the lack of knowledge of how to develop them, which is the primary issue addressed in this work. However, the problems do not stop after an MPCS has been designed. The introduction of MPCSs, or the modification of existing MPCSs, requires carefully selected migration strategies to avoid any deterioration of the manufacturing performance during the modification. We will discuss some important aspects of the application and introduction.

1.5 Overview of this Book

Introduction. In Part I, we define an MPCS and develop a systematic and methodical approach to design an integrated and flexible MPCSs.

It appears that an MPCS is a very complex system. To master the complexity, we have to develop several specifications of an MPCS. We start at a high level of abstraction, where we describe the functionality of MPCS components, and step by step replace the abstract specifications by more implementation oriented specifications.

Moreover, we have to identify the components of an MPCS and determine their tasks. To that end, we use decomposition methods: we first consider an MPCS as a black box and, in several steps, decompose this black box into a configuration of MPCS components. This configuration of components replaces the black box.

We will decompose an MPCS to develop a configuration of MPCS components and give an abstract specifications of this configuration. We call this specification a 'Reference Model'.

A 'Reference Model' describes a complex system, such as an MPCS, as a configuration of components that each execute their own, globally defined, distinct tasks but interact to realise the task of the system as a whole.

We will develop two Reference Models, i.e. a 'Reference Model for MPCSs' and a 'Reference Model for MPCS Management'.

The Reference Model for MPCSs describes an inflexible MPCS, which operates in a stable environment: it does not interact with the environment to change its product portfolio, production capacity, or production costs. In case of a flexible MPCS, however, these all change gradually. We therefore propose a system, 'MPCS Management', that can prepare an MPCS for these changes by feeding it with appropriate information about the changes so as to realise a flexible MPCS.

Reference Model for MPCSs.   In Part II, we develop a Reference Model for MPCSs. Step by step, we decompose the black box description of the MPCS. In each step, we identify components, which we may have to decompose further in a next step, define the tasks and interactions of the newly-identified components, and justify the decomposition. Moreover, we describe real-life scenarios and describe how our MPCS can play a role in these scenarios.

The resulting Reference Model describes the MPCS as an organisation of components with such tasks as:

- the control of inventory levels to be able to quickly dispatch products while keeping inventory costs low;

- the scheduling of operations;

- the coordination of machines to execute operations;

- the determination of the trajectories of joints of machines;

- the servoing of joints; and

- related feed back tasks.

Reference Model for MPCS Management. In Part III, we discuss MPCS Management in detail and propose a Reference Model for MPCS Management. We first describe the management activities, such as changing the product portfolio and preparing the MPCS for new production targets. We then decompose MPCS Management into a configuration of components, using the decomposition techniques discussed in Part I.

The resulting Reference Model describes MPCS Management as an organisation of components with such tasks as:

- analysis of the feasibility of production targets;

- product design;

- machine design;

- process planning;

- development of control procedures;

- maintenance;

- monitoring, etc.

Conclusions. In Part IV, we elaborate on the major conclusion, i.e. that we have contributed to the development of a systematic and methodical approach to manufacturing control by developing Reference Models for MPCSs and MPCS Management.

Both, the Reference Models for MPCSs and MPCS Management, are indispensable mile stones in developing an integrated MPCS. They define components of the total, complex MPCS, and meanwhile allow us to keep sight of the relationships between the separate components and between the components and the MPCS as a whole.

The Reference Models are also indispensable mile stones on the way towards flexible MPCSs. The models are based on the analysis of a wide variety of practical applications and are therefore considered valid for these applications. Moreover, MPCS Management, which serves to realise general applicability, has been analysed in considerable detail.

In Part IV, we also discuss the practical use and application of the theories and concepts developed in this work and discuss directions for future research.

Uniqueness of the Designed MPCS. We develop and justify a design of an MPCS that has evolved from an endless number of micro decisions and trade-offs, and is subject to private judgement and taste. It is a satisfactory, workable design, justified on the basis of sound design heuristics but by no means can it guarantee a unique solution for an integrated and flexible MPCS.

Intended Readers. There are three requisites for appreciating this work. Firstly, the need for top-down design techniques in the development of complex systems should be appreciated. Secondly, a familiarity with MPCSs and an ability to relate particular observations to the more general and abstract descriptions in this work would be helpful. Thirdly, and most importantly, a curiosity to see a 'bigger picture' is essential, a picture which includes problems of many domains which are discussed in isolation by typical studies or text books.

This document describes the MPCS in detail. For brief papers that summarise the essence of this book, refer to [22,23,24].

Note. We have been intensely involved in the development and standardisation of Reference Models in a company, which led to the wide-spread publication of an internal standard for a Reference Model [8,9]. This work gives a scientific justification of this standard but is wider in scope since it includes MPCS Management. It adheres as much as possible to the terminology in [8,9], without justifying or motivating it.

Chapter 2

Manufacturing Planning and Control Systems

In this chapter, we describe a 'Manufacturing Planning and Control System', or 'MPCS'. An MPCS executes the manufacturing planning and control tasks of a 'Production Organisation'. We discuss a Production Organisation first.

2.1   The Context

A 'Production Organisation' is a system that can manufacture products in accordance to prescribed production targets, provided that it is supplied with the resources needed to manufacture these products.

A 'Product' is an end-item for manufacturing and a commodity for sale.

'Resources' are non-renewable items such as raw materials or renewable items such as tools, product designs, and machines.

Figure 2.1 gives a simplified illustration of a Production Organisation: it receives production targets from a 'Company Controller', receives resources from 'Suppliers', and dispatches products to 'Customers'.

Company Controller.   We do not define a Company Controller in detail, but focus on the fact that it is a system that pursues profits by the manufacture and sales of products.   As Figure 2.2 illustrates, it uses the services of support organisations such as 'Production Organisations', 'Marketing Organisations', and possibly others. Marketing Organisations identify and explore opportunities to market products. Pro-

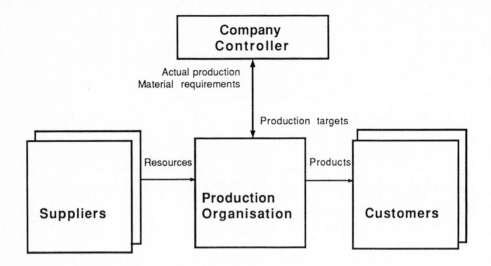

Figure 2.1: A Production Organisation Interacting With Suppliers, Customers, And Company Controller

duction Organisations manufacture those products. Marketing Organisations may, but not necessarily have to, act as Customers for a Production Organisation.

The Company Controller should define sales and production targets for its Marketing and Production Organisations respectively. The production and sales targets should be coherent, serve the Company Controller's interest, and be feasible for the Production and Marketing Organisations. The Company Controller may give definite targets for the immediate planning horizon but may also provide forecast targets for future horizons, which allows the support organisations to prepare for those targets.

A Production Organisation reports to the Company Controller its material requirements, i.e. which materials it needs to realise its production targets. Further, it reports about the actual production, for example, to notify the Company Controller that a production target has been realised.

In fact, the Company Controller and Production Organisation negotiate the production targets since they should be commercially viable for the Company Controller and feasible and cost-effective for the Production Organisation. The Company Controller may have various Production Organisations that can manufacture the same products. Although not shown in Figure 2.2, a Production Organisation should report its manufacturing costs so that these can be taken into account when the Company Controller allocates the production targets. The more cost-effective Production Organisations

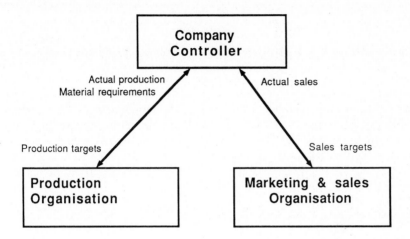

Figure 2.2: The Company Controller And Support Organisations

have a better chance to receive production targets, which is an added incentive for them to improve the cost/performance ratio.

Production Organisation as Cost Centre. A Production Organisation is a 'cost centre': it is not concerned with making profit but with executing a certain task, i.e. manufacturing, at certain agreed upon costs. A Production Organisation is therefore not concerned with decisions as to whether investments in production, marketing, or sales are needed. Neither is it concerned with the selection of Customers and Suppliers.

The Company Controller is responsible for the profits and therefore is concerned with external factors that impact the profits, such as the selection of targeted Customers and Suppliers. It negotiates the sales of products with Customers of its Production Organisations (which may be a Marketing organisation), agrees with the Production Organisations on their production targets, and negotiates with Suppliers the delivery of resources needed by the Production Organisations.

Let us now consider a few representative scenarios that capture the majority of real-life Production Organisations and show that their basic task of realising production targets as agreed upon with the Company Controller is a purposeful one.

'Make-to-Stock'.   'Make-to-Stock' commonly refers to situations where the Company Controller sets firm production targets for the Production Organisation before it has received orders from external customers for the products involved.

Typically, companies that operate in a make-to-stock mode, have sales and distribution organisations that accept products from the Production Organisations, transport them, and store them in warehouses. They ship the products to external customers, such as dealers and retail stores, as soon as these order them. Fluctuations in customer demand are buffered by the stocks in the warehouses and do not immediately influence the dispatch rate of the Production Organisations. As a result, the Company Controller can accept relatively large time slots for the realisation of production targets when it negotiates these with the Production Organisation. It is up to the Production Organisation to determine when in these time slots on which it actually dispatches the products.

The Marketing Organisations report to the Company Controller about the actual or expected demand for products. The Company Controller determines which markets should be pursued and gives sales targets to the Marketing Organisations. It should ensure that these Marketing Organisations, which act as Customers for the Production Organisations, receive the products to realise their sales targets. The Company Controller therefore negotiates with the Production Organisations to assess whether they can manufacture the products required.

A Production Organisation reports to the Company Controller which raw materials it needs and when it can dispatch the products. The Company Controller negotiates the delivery of the raw materials with Suppliers.

The process of analysing the sales forecasts and selecting the production targets is repeated every month or so. The production targets define the required output of products for the next month. Typically, forecast production targets are given well in advance to enable the Production Organisation to develop manufacturing capabilities or to enable it to acquire raw materials.

The Company Controller can ensure a reliable and cost-effective supply of raw materials in several ways. The Company Controller can, for example, negotiate long-term contracts for the supply of raw materials. The Company Controller may also procure materials that are not yet needed by the Production Organisations but will be needed in the future, in order to maintain good relations with the Suppliers. The Company Controller may also consolidate the material requirements of several of its Production Organisations to profit from quantity discounts or better utilise carriers that transport the resources. Finally, the Company Controller may actually control some of the suppliers.

Example.   A Company Controller controls three Production Organisations, one that makes glass, one that makes fluorescent lamps, and one that makes incandescent lamps. The glass producer supplies glass to the fluorescent lamps producer and to the incandescent lamp producer, besides the external market.

We described Production Organisations in a make-to-stock situation. We will now discuss Production Organisations in a make-to-order situation.

Make-to-Order.   'Make-to-Order' commonly refers to situations where the Company Controller gives firm production targets to the Production Organisation before it has received customer orders for the products involved. Typically, the Production Organisation has to realise the orders relatively quickly since the customer expects the products. The Company Controller therefore defines narrow time slots on short notice and the Production Organisation is required to dispatch certain products in these time slots. Instead of giving targets that should be realised somewhere in, say, the next month, as in make-to-stock situations, the Company Controller determines which products should be dispatched at, say, a specific day.

The Company Controller uses the sales forecasts of the Marketing Organisations to select markets and to determine the required production capability and capacity of its Production Organisations.

The Company Controller negotiates with Customers the price, due dates, location of delivery of products. However, it has to negotiate with the Production Organisations whether the products can be delivered in time at acceptable costs. The Production Organisation may not be able to supply the products in time, could supply them contingent on the availability of certain raw materials, prefers to supply them in batches, may need time to develop new products, etc. The Company Controller tries to reconcile the customer demand and the Production Organisation's capacity. To that effect, it may propose discounts on quantum sales, or on sales with convenient due dates, etc. The Company Controller further has to negotiate with suppliers so that the Production Organisations receive the resources they need. The costs and availability of resources affect the costs and potential due dates of the products.

Comparing Production-Organisations in a make-to-stock and make-to-order situation, we may conclude that the Company Controller executes much more stringent control in a make-to-order situation. However, our global description of a Product

Organisation, as illustrated in Figure 2.1 can apply to make-to-stock as well as to make-to-order situations.

In the following chapters, we refine our description of a Production Organisation. We will describe, for example, how a Production Organisation can react to forecast production targets and to requests from the Company Controller to change its product portfolio, production capacity, or production costs. We will also describe how the Production Organisation and Company Controller negotiate budgets. Further, we develop an internal organisation for the Production Organisation that shows *how* it manages to realise production targets. This internal organisation should apply to a wide variety of applications, including make-to-stock and make-to-order applications.

Level of Abstraction. Our treatment of Production Organisations is intentionally abstract. We focus on their role and functionality in an environment consisting of Company Controller, Customers, and Suppliers; we abstract from their physical implementation. The reason is that we have to define role and function before we can design physical implementations.

It is important to keep this in mind when comparing our descriptions with real-life observations. We have, for example, assigned the purchasing tasks to the Company Controller on the basis of its responsibility to select Customers and Suppliers so as to maximise its profits and on the basis of its access to information about market demand, production costs, and raw material costs. However, often the real-life Production Organisations reporting to the same Company Controller have a local purchasing department 'within their walls'.

It looks therefore as if purchasing is a task for a Production Organisation rather than for the Company Controller. However, we should distinguish between the abstract and physical Production Organisation and Company Controller.

The purchasing task is very much a Company Controller task. However, for practical reasons, a portion of the purchasing task is often executed by a 'local' department, people within the physical system that implements the Production Organisation. A 'central' department, people within the physical system that implements the Company Controller, executes the remaining purchasing tasks. There are some reasons why, in specific situations, such implementations with a local purchasing department make sense:

- the local purchasing in effect calls from Suppliers to supply materials according to the long term agreements made by these Suppliers and the central Company Controller. Thus, local purchasing merely executes the long term decisions made by the central purchasing group. Due to its proximity to manufacturing, it knows when the materials are needed.

- there would not be a significant difference in the costs of raw materials if they were purchased by local or central purchasing departments. The argument that purchasing should be centralised to optimise profit does not hold in this case. The profits are rather insensitive to the prices of raw materials.

- the raw materials are only needed by one Production Organisation. There is no need for a coordinated purchasing by a central Company Controller;

- the Suppliers are at short distance, work the same hours as, and speak the same language as the local purchasers. The local purchasers are in a better position than the central purchasers to deal with the local Suppliers.

In all these situations, the local purchasing activities can be viewed as Company Controller tasks executed by people within the same building as the people that execute Production Organisation tasks.

The practical situation is a violation of our abstract model if the local purchasing tasks are considered Production Organisation tasks. This happens, for example, if the costs of the local purchasing activities are accrued to the real-life Production Organisation instead of the Company Controller. In that case, the Company Controller gives budgets to the Production Organisation for manufacturing costs, but portions of the budget are used to do tasks that essentially are Company Controller tasks. This prevents the Company Controller from getting an accurate assessment of the Production Organisation as a cost centre for manufacturing. Admittedly, this is often the case. It illustrates one difference between practice and what might be a more ideal situation.

Manufacturing Planning and Control System. We investigate the Production Organisation, and in particular how it plans and controls the activities of sensors and actuators on the shop floor so that it realises its production targets. We focus on the manufacturing planning and control aspects and ignore other aspects in a Production Organisation relating to personnel, accounting, administration, safety, cleaning, etc.

We therefore refer to the system in a Production Organisation responsible for the manufacturing planning or control as a 'Manufacturing Planning and Control System', or 'MPCS'.

We will describe MPCSs extensively throughout this document. Let us first illustrate the global task of an MPCS by specifying an instance of an MPCS that executes this task.

2.2   An Informal Specification as an Example

Once we know the global task of a system, such as an MPCS, we can completely specify its behaviour by defining its interactions with its environment, i.e. which information it exchanges, when, and under which conditions.

For example, we can specify that a specific MPCS accepts certain production targets, but only after it has told the Company Controller that previously received production targets have been realised. Further, this MPCS will dispatch products provided that it has received a production target outlining those products and provided that it has received the appropriate raw materials.

One could think of a variety of such specifications that would all satisfy the global task description of an MPCS as given in Section 2.1. The suitability of a specific specification of a system depends on the envisaged application of the system.

We refer to the information exchanges of systems as prescribed by their specifications as 'interactions'. The interaction concept is an advanced concept that replaces the traditional concept of input and output [31]. An interaction can be considered as a common activity of two or more interacting processes in which information is established to which these processes can refer. Interactions can be described by interaction primitives.

Interaction Primitives.

   'Interaction primitives' define units of information exchanged by interacting systems, without reference to the physical implementation of the information or the mechanisms of information exchange.

Appendix A describes the meaning and purpose of interaction primitives in some detail. For now, it is sufficient to think of them as a set of variables that get certain values to be agreed upon by and accessible to the systems that participate in the interaction. Take, for example, the Production Target interaction primitive, defined in Table 2.1. It specifies that both the Company Controller and MPCS agree on the products and raw materials to be dispatched and received by the MPCS in terms of the number and type of products and raw materials, the Customers and Suppliers, and the location and date of product and raw material exchanges.

An interaction primitive does not specify how the information is generated. The Production Target interaction primitive, for example, does not tell whether the Company Controller, or MPCS, or both propose the products to be produced. An interaction that takes place, therefore implies a kind of negotiation, which is not explicitly described, to establish mutually agreeable parameter values. The Production Target interaction primitive, for example, does not reveal potential negotiations of the Company Controller and MPCS that may in reality take place to select acceptable production targets; it only describes the end result, i.e. which production target has been established.

Note that we could replace the Production Target interaction primitive with primitives that explicitly define proposals and final agreements in the negotiations if we were interested in modeling such negotiations. In this way, we can refine the specifications step by step: we start with abstract interaction primitives and replace them if we want to develop more detailed specifications.

Specifications as Illustrations of Task Definitions. We will give specifications of MPCS components to *illustrate* their global task definitions. We do not intend to develop complete or generally-applicable specifications of MPCS components. In fact, we will primarily give examples of the interactions, and seldom describe the timing of the interactions. Furthermore, we do not discuss refined specifications. The primitives in Tables 2.1 and 2.2, for example, suffice to convey the essence of the relation of an MPCS and its environment, i.e. that Company Controller and MPCS agree upon production targets, that Customers, Suppliers, and MPCS exchange products.

Interaction Primitives for a Specific MPCS. The interaction primitives in Tables 2.1 and 2.2 can be used to model an MPCS that:

- Agrees with the Company Controller on a production target provided that it receives certain raw materials. The Product parameters can identify unique product types, or groups of products types. They can also identify a certain number of products.

  According to the Production Target primitive proposed in Table 2.1, the MPCS may require that some raw materials be supplied. The Company Controller and MPCS agree, for example, that the MPCS dispatches certain television sets contingent upon its reception of tubes and other raw materials.

  Note that an MPCS does not necessarily ask for raw materials that are needed to dispatch the products mentioned in the same Production Target. It could, for example, agree to dispatch products that it has in stock already. It does not need raw materials to dispatch these products. However, it may require raw materials to be prepared for future production targets, and can establish this requirement with the Production Target interaction primitive.

- Agrees with the Company Controller to prepare for forecast production targets.

- Provides the Company Controller with status about the realisation of production targets by the Production Indication primitive, which describes which raw materials and products have been exchanged, at which location and time in order to realise a certain production target. Note that products may be exchanged that differ from those defined in the production target. The MPCS may, for example, indicate that it has received defective raw materials. Several progress parameters can be defined to indicate progress is real or expected, that progress is blocked due to material shortage, etc.

- Can indicate that it needs raw materials, independent of a specific production target, for example, to compensate for wasted materials.

- Can exchange raw materials and products with Product Exchange primitives. If we speak of 'exchanging' products, we actually mean the exchange of the *control* of products, which does not necessarily imply a physical transfer of products.

Interaction Primitive	Arguments	Explanation
Production-Target	Products	Target to ship *products*
	Partner	to a *customer*
	Location	at a *location*
	Due-date	and a *due date*
	Products	and to accept *products*
	Partner	from a *supplier*
	Location	at a *location*
	Due-date	and a *due date.*
	Command-Id	*Identification* for further reference.
Production-Target-Forecast	Products	Forecast of a target to ship *products*
	Partner	to a *customer*
	Location	at a *location*
	Due-date	and a *due date*
	Products	and to accept *products*
	Partner	from a *supplier*
	Location	at a *location*
	Due-date	and a *due date.*
Production-Indication	Products	Report that *products*
	Partner	have been shipped to a *customer*
	Location	at a *location*
	Due-date	and a *due date*
	Products	and that *products*
	partner	from a *supplier*
	Location	at a *location*
	Due-date	and a *due date.*
	Progress	Information about the *progress.*
	Command-Id	*Reference* to production target.
Raw Material-Request	Products	Request to acquire *raw materials*
	Location	at a *location*
	Due-date	and a *due date.*

Table 2.1: Interactions Of Company Controller And MPCS

26

Interaction Primitive	Arguments	Explanation
Product-Exchange	Products Sender Receiver	Exchange of *products* from a *sender* to a *receiver*.

Table 2.2: Interactions Of Customers, Suppliers, And MPCS

Chapter 3

Development of MPCSs

This chapter explains how we propose to develop an integrated and flexible MPCS.

3.1   Top Down Approach

The complexity and uncertainty pertaining to the MPCS itself as well as to the process to develop an MPCS have a considerable impact on the approach we choose to develop an MPCS. We discuss the complexity and uncertainty pertaining to the MPCS first.

Complexity of an MPCS.   The problems to be solved by an MPCS to realise its production targets are formidably complex. Consider, for example, the problems of scheduling jobs on machines or planning the path for a robot arm. Due to the combinatorial nature, these tasks soon become computationally intractable if the dimension of the search space is extended, and particularly so if exhaustive evaluations of all possible solutions are required [45,83]. There is no way that realistic investments in computational power can remedy the intractability of algorithms that require exhaustive search [30].

Apart from the exceedingly large numbers of non-dominant solutions, it is often difficult to clearly quantify the value of a given solution. This makes it difficult to assess solutions which have been generated. How to assess, for example, the decision of an MPCS to request raw materials to prepare for the manufacturing of products according to forecasts production targets, if finally the forecasts do not materialise?

In light of the intractable nature of computations required in an MPCS, we have to focus on methods to compute a *satisfactory* way of achieving a goal with a reasonable investment in computational power and time [50,76,96]. The computational goal is not specified as an optimum to be found, but rather as a range of goals, which are equally desirable. The fact that a satisfactory solution is accepted implies that optimal solutions are pursued only if justified by reasonable computational costs [1].

Uncertainty facing an MPCS. Another source of complexity is the uncertainty and unpredictability of the environment of an MPCS [45] with its changing product portfolios, its changing requirements regarding product portfolios production capacity, and production costs.

An MPCS component performs computations to act upon some part of its environment. Plans are made to that effect predictively without advance knowledge of how the environment will react. If response is not as expected, plans may no longer be feasible and will have to be generated again.

The question is not whether the environment will respond as expected but whether the MPCS can adapt to changes in the environment. The environment does change: production targets may vary widely in mix and volume and call for new products, the life cycles of products become shorter, the capacity of an MPCS has to be expanded or reduced, machines have to be replaced with more advanced ones, etc.

Since the computations of the MPCS determine how it interacts with their environment, changes in the environment will have consequential effects on the computations required. For practical reasons, it is imperative that the MPCS be easily adaptable to such changes.

Impact on Development. We have discussed how the MPCS's inherent complexity and the uncertainty of its environment imply that the MPCS should pursue satisfactory performance for its manufacturing planning and control task and that the MPCS be easily adaptable to changes in its environment. Complexity and uncertainty therefore affect the MPCS, which after all is a desired result of our development process. However, these two phenomena also affect the development process itself.

Complexity of the Development Process. The complexity of an MPCS forces consideration of MPCS designs at various levels of abstraction so that we can ignore details at each level that can be considered separately. In a series of steps, specifications at successively lower levels of abstraction implement those that are made

---

[1]May be, we should view satisfactory design *solutions* as optimal, if we include the costs of computation as a parameter of the function to be optimised.

at higher levels.

At the highest level of abstraction, we focus, for example, on the global task of a component. At a lower level of abstraction, we define the relation of the inputs and outputs of the component so that it realises its global task. At an even lower level of abstraction, we focus on the physical implementation of the component.

The complexity of an MPCS also influences the choice of which and how many abstraction levels should be distinguished in a particular design process. This affects the difference in abstraction levels that a designer can successfully bridge.

As a result of the step-by-step approach, optimality of design cannot be guaranteed. It is difficult to accurately foresee, for example, the consequences of a high-level design decision, such as the desired functionality, for a low level design decision such as the costs of material of a product with this functionality. It is, therefore, not possible to decide beforehand whether the desired functionality is worth the costs. The design process, therefore, aims at a satisfactory, rather than optimal design [2].

Uncertainty affecting the Development Process. There will always be uncertainty as to whether a system can still be usefully applied when its design is finished, as anticipated at the beginning of its design. Similarly, there is uncertainty about constraints that play a role during the design process such as the costs of material that can be used, for example.

The uncertainty, just as the complexity, affects our selection of abstraction levels. We want to minimise the likelihood that designs have to be changed as a result of the changes mentioned. The generality of a design, the ability to use it as a starting point for many implementation processes, is an important criterion for the selection of the levels of abstraction.

> **Example.** MPCS components have to cooperate, but they appear in different configurations and different implementation agencies determine the actual physical construction of the different components. Designs that describe the cooperation of components, independent of their configuration or physical construction, will be applicable irrespective of the implementation agencies involved [101].

Consequences of Complexity and Uncertainty. We have described that the

---

[2]May be, we should view the satisfactory design *process* as optimal if we consider the amount of computation needed by the designer to produce a design as a parameter in the cost function.

complexity and uncertainty affect the way the MPCS executes its task as well as the way we should develop an MPCS. We conclude that:

- an MPCS is sufficiently complex that we have to accept a design that realises an MPCS with satisfactory performance;

- the environment of an MPCS may require significant changes of the MPCS so that we should develop an MPCS that is flexible to adapt to these changes;

- an MPCS is sufficiently complex that we have to develop it step by step, starting with abstract designs that allow us to understand the MPCS despite its complexity; and

- physical MPCSs vary in many respects so that we should pursue generally applicable intermediate design results that can be used for the development of many MPCSs.

In the next Sections, we will discuss a series of development steps that meet these requirements. We start with the description of an MPCS as a black box. Subsequently, we develop designs of an MPCS at lower levels of abstraction, which implement the black box, till ultimately a physical MPCS results.

3.2 Inflexible MPCS, with Stable Product Portfolio, Production Capacity, and Production Costs

To deal with the formidable complexity of a flexible MPCS, we first make a simplifying assumption, i.e. that the MPCS will not engage in interactions with its environment that would require it to change its physical implementation. For example, we assume that the MPCS can manufacture certain products, but will not be required to manufacture new products since that could also imply that it would need new machines. Similarly, we assume that it can produce products in certain volumes, but is not required to increase its capacity since that could imply the development of new machines as well.

More specifically, we abstract from possible interactions of the MPCS and Company Controller to change its product portfolio, production capacity, and production costs.

The 'product portfolio' of an MPCS is the range of products that an MPCS can manufacture and dispatch.

The 'production capacity' of an MPCS is a measure for the size of the production targets that it can realise, typically expressed in terms of the numbers of units of prod-

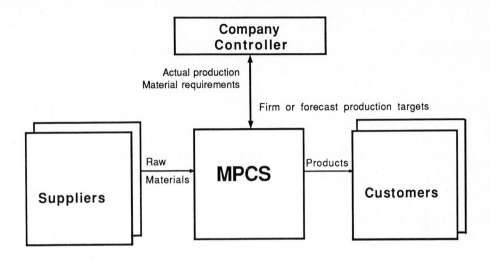

Figure 3.1: Inflexible MPCS And Its Environment

ucts produced in unit time and in the response time to dispatch a given quantity of products.

The 'production costs' of an MPCS are the costs incurred by an MPCS to transform specific raw materials into specific products in specific volumes.

In other words, we assume that the interactions of the MPCS and its environment do not require it to *change* its 'application'.

The 'application of an MPCS' defines its current product portfolio, production capacity, and production costs.

An 'inflexible MPCS' is an MPCS that cannot change its application.

Figure 3.1 illustrates an inflexible MPCS; there are no interactions of MPCS and Company Controller regarding modifications of its application.

As a result of our simplifying assumption, we abstract from the mechanisms that would enable an MPCS to change its application: the design of products, design of machines and their layout, the development of scheduling algorithms, hiring or firing personnel, etc. Our inflexible MPCS is 'only' concerned with the planning and control of an existing manufacturing facility to manufacture existing products.

In Part II of this work, we will investigate, how we can develop an inflexible MPCS. In part III, we will investigate how we can develop a flexible MPCS, which is capable of changing its application.

3.2.1 Global Tasks of a Black Box MPCS

We start our investigation by describing an inflexible MPCS as a black box. This is the perspective of an MPCS from its environment, i.e. the Company Controller, Customers, and Suppliers. As is illustrated in Figure 3.2, the black box description is a starting point for the design of an integrated MPCS.

In chapter 2, we described the MPCS as a black box by describing its global task.

A 'global task' characterises a task without completely specifying the task.

On the basis of a system's global task definition, we can develop the complete functional definition of the behaviour of the system.

The 'complete functional behaviour' of a system defines the system, without reference to its physical implementation, in terms of its possible interactions with its environment by giving precise definitions of the interactions and their temporal ordering.

To illustrate the difference between 'global task' and 'complete functional behaviour', consider the MPCS. The definition of its global task includes accepting production targets, accepting raw materials, and dispatching products according to these targets. If we were to define its complete functional behaviour, we would have to specify in a precise way:

- the interactions of the MPCS and the Company Controller to establish the production targets;

- the interactions of the MPCS and Customers and Suppliers to exchange the raw materials and products respectively; and

- the relations of these interactions in terms of their contents and temporal ordering.

Black Box as Basis for Complete Functional Specifications. We could propose a specification of the complete functional behaviour of the black box MPCS on the basis of its global task definition. The specification would have to define an MPCS that executes the global task defined above. However,

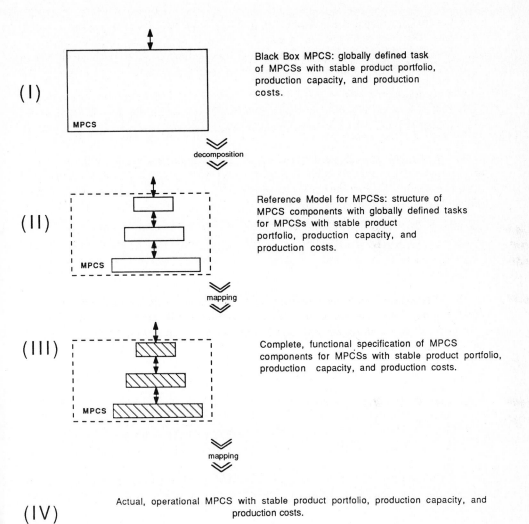

(I) Black Box MPCS: globally defined task of MPCSs with stable product portfolio, production capacity, and production costs.

MPCS

decomposition

(II) Reference Model for MPCSs: structure of MPCS components with globally defined tasks for MPCSs with stable product portfolio, production capacity, and production costs.

MPCS

mapping

(III) Complete, functional specification of MPCS components for MPCSs with stable product portfolio, production capacity, and production costs.

MPCS

mapping

(IV) Actual, operational MPCS with stable product portfolio, production capacity, and production costs.

Figure 3.2: Steps In The Design Of An Inflexible MPCS

- developing such a specification requires the selection of the scenarios in which the MPCS must be able to operate. The complete functional specifications are less generally applicable than the global task definitions: the specification would render an MPCS that can only operate in specific scenarios.

- a complete functional specification of a black box MPCS would describe a system that is way too complex for a physical implementation. The specification would therefore be of little use to people who want to design and build MPCSs.

We should first develop a design of an MPCS that is more widely applicable than the complete functional specification of the black box and that facilitates a physical implementation. We therefore propose to defer the specification of the complete functional behaviour of the black box MPCS. We should first decompose the black box MPCS into a structure of components with globally defined tasks.

Decomposition.

'Decomposition' of a system represented as a black box with a globally defined task, is the replacement of this representation with a configuration of system components, each with its own globally defined subtask, that interact to realise the globally defined task of the black box.

Thus, a decomposition of the black box MPCS should result in MPCS components:

- with globally defined tasks; and

- that execute relatively simple tasks so that the complete functional specifications of these components, if so developed, are suitable for physical implementation.

We will therefore decompose the black box MPCS on the basis of its global task definition. The decomposition is illustrated by the transition from Section I to Section II in Figure 3.2.

The result of the decomposition is a 'Reference Model', as discussed in the next Section.

3.2.2   Reference Model for MPCSs

We could decompose an MPCS in many ways but we are only interested in decompositions that meet an important criterion. We require that the MPCS components execute relatively simple tasks. The components execute simple tasks if they can be viewed and understood as distinct entities in the context of the overall system.

The MPCS should therefore be decomposed into components that interact sparingly to reconcile the fact that they interact and the requirement that they can be viewed separately in the context of the overall system. In other words, the components should be relatively independent because the state transitions due to their own, internal processing significantly outnumber those due to their interactions with other components.

We call the result of a decomposition of a black box system that consists of relatively independent components a 'Reference Model'. More precisely:

A 'Reference Model' represents a system as an organisation in terms of a structure of relatively independent, interacting components, and in terms of the globally defined tasks of these components.

Note that a Reference Model, defining tasks, does not imply any reference to any physical means with which the model may be implemented. For the sake of clarity, Appendix D contains some divergent interpretations of 'Reference Model' that do not conform to the definition used in this work.

Although not shown in Figure 3.2, it will appear that a sequence of decomposition steps can be carried out to develop the Reference Model for MPCSs. In a first step, the MPCS is decomposed into a few components. In a next step, these components can be decompose further, etc. We will develop a Reference Model for MPCSs in Part II of this work.

3.2.3 Complete Functional Specification of MPCS Components

We can specify the complete functionality of each component so that it fulfills the requirements of its globally defined task as defined by the Reference Model. To completely specify a component, we define its allowable interactions and their relation with respect to contents and temporal ordering.

We do not yet define the physical representation of the interactions since no decision has been made yet on the distribution of components over physical systems.

Sections II and III in Figure 3.2 illustrate that the complete functional behaviour of MPCS components can be specified using their global task definitions as a starting point.

Mapping. Starting with the Reference Model of an MPCS, we obtain the complete functional specifications by means of 'mapping'.

Mapping is different from decomposition, as follows from the following definition.

'Mapping' is the replacement of a representation of a system, either as a black box or as a configuration of components, that interacts in a certain way with its environment, with a representation of a system that interacts in a different way with the environment to realise the task of the first system; explicit rules prescribe how the interactions of the latter system should be interpreted in terms of the first system.

The mapping of the Reference Model onto the complete functional specifications of MPCS components implies that the specifications should describe MPCS components that interact with their environment to execute their global task as defined by the Reference Model.

Furthermore, the specifications should define MPCS components that interact to realise a behaviour that, from the perspective of an external observer, is equivalent to the complete functional behaviour of the black box MPCS.

3.2.4   Physical Implementation

In a next step, illustrated by the transition from Section III to IV in Figure 3.2, we should determine how to physically implement the components.

The interaction primitives of the complete functional specification of MPCS components have to be mapped onto physical interactions. The resulting specifications of physical components can be given to such different implementation agencies as engineering groups, software houses, and vendors. Multiple implementation agencies can therefore contribute to the construction of an MPCS.

3.3   Flexible MPCS, with Modifiable Product Portfolio, Production Capacity, and Production Costs

In the previous Sections, we reviewed a strategy to develop an MPCS. We made the simplifying assumption that the MPCS does not have to modify its product portfolio, production capacity, or production costs. Such MPCSs, with a stable application, exist. However, the vast majority of MPCSs is required to change its application regularly.

Consequently, it makes sense to investigate the development of an MPCS that can change its application. Moreover, we stated explicitly in the introductory text of chapter 1 that we pursue flexible MPCSs.

A 'flexible MPCS' is an MPCS that can honour requests for gradual changes in its current application.

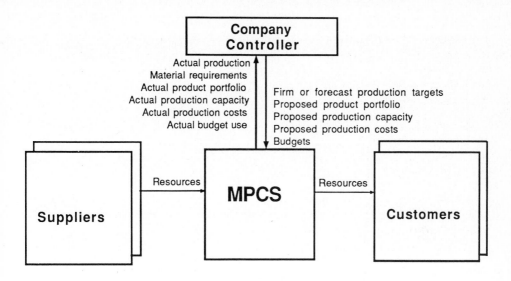

Figure 3.3: A Flexible MPCS And Its Environment

Changes in the application will be gradually in the sense that they occur less frequently than production targets are established. Figure 3.3 illustrates a flexible MPCS. Unlike the inflexible MPCS, illustrated in Figure 3.1, the flexible MPCS does interact with the Company Controller to change its application.

We will now discuss a method to develop a flexible MPCS. This method will be applied in Part III of this work. It will appear that the design of an inflexible MPCS, which is developed in part II, is of use for the development of a flexible MPCS [3].

We replace the black box description of an inflexible MPCS with a black box description of a flexible MPCS. Section I in Figure 3.4 illustrates the black box description of a flexible MPCS.

The development strategy reviewed in Sections 3.2.1 through 3.2.4 need to be extended to develop a flexible MPCS. The reason is that the complete functional specifications developed on the basis of global task definitions of the Reference Model would specify components for specific applications. They would, for example, specify MPCS components that can contribute to the manufacturing of only a limited number of products.

---

[3] Readers could choose to defer reading of this Section till they have finished Part II.

38

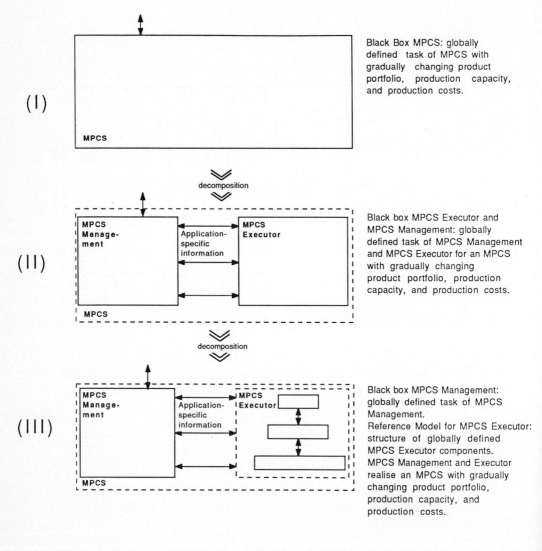

(I)

Black Box MPCS: globally
defined task of MPCS with
gradually changing product
portfolio, production capacity,
and production costs.

decomposition

(II)

Black box MPCS Executor and
MPCS Management: globally
defined task of MPCS Management
and MPCS Executor for an MPCS
with gradually changing
product portfolio, production
capacity, and production costs.

decomposition

(III)

Black box MPCS Management:
globally defined task of MPCS
Management.
Reference Model for MPCS Executor:
structure of globally defined
MPCS Executor components.
MPCS Management and Executor
realise an MPCS with gradually
changing product portfolio,
production capacity, and
production costs.

Figure 3.4: Steps In The design Of A Flexible MPCS

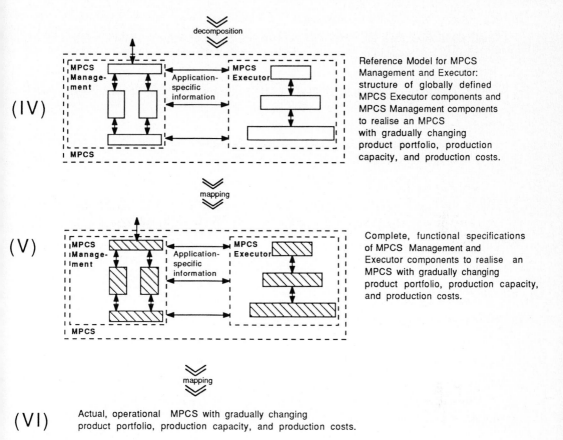

(IV)

Reference Model for MPCS
Management and Executor:
structure of globally defined
MPCS Executor components and
MPCS Management components
to realise an MPCS
with gradually changing
product portfolio, production
capacity, and production costs.

(V)

Complete, functional specifications
of MPCS Management and
Executor components to realise an
MPCS with gradually changing
product portfolio, production capacity,
and production costs.

(VI) Actual, operational MPCS with gradually changing
product portfolio, production capacity, and production costs.

Figure 3.4: (continued)

Rather than specific MPCS components, we now need generally applicable MPCS components that can be instantiated, i.e. components that can operate in a specific application after they have received information about the specific application. Think of information describing the products to be manufactured, or control procedures to be executed, etc.

We therefore implement the flexible MPCS by two system components, i.e. one that:

- can execute the MPCS tasks in a certain application provided that it has received application-specific information; and one that

- defines and installs the application-specific information.

We call these systems 'MPCS Executor' and 'MPCS Management' respectively.

3.3.1   MPCS Management and Execution

An 'MPCS Executor' is a system that can behave as an inflexible MPCS provided that it is supplied with application-specific information and mechanisms.

'MPCS Management' is a system that can provide an MPCS Executor with application-specific information and mechanisms to prepare it for operation in its designated application.

We use 'mechanism' here to denote machines, tools, physical MPCS Executor components, etc.

We decompose the flexible black box MPCS (Section I of Figure 3.4) into MPCS Management and MPCS Executor. The above definitions give the global task definitions of the black boxes MPCS Management and MPCS Execution that result from this decomposition.

We can now repeat arguments made before and conclude that black boxes MPCS Management and MPCS Executor need to be decomposed.

Decomposition of MPCS Executor.   We need to decompose the MPCS Executor to develop a Reference Model for the MPCS Executor. This decomposition, illustrated by the transition from Section II to III in Figure 3.4 is relatively easy. As we will discuss in more detail in Part III of this work, it makes sense to give the MPCS Executor a structure that reflects the structure of MPCS components by mapping the components of the Reference Model for MPCSs, developed in Part II of this work, onto the MPCS Executor components on a component-by-component basis. MPCS Executor components, when loaded with the application-specific information, behave as the corresponding components of the Reference Model for MPCSs. They will execute relatively independent tasks since we gave the components of the Reference Model relatively independent tasks. The decomposition of the black box MPCS Executor will therefore meet the requirement, explained in Section 3.2.2, that the components of a Reference Model be relatively independent.

Section III of Figure 3.4 still represents MPCS Management as a black box; we cannot immediately reveal an internal organisation.

3.3.2 Reference Model for MPCS Management

We decompose the black box MPCS Management, like we decomposed the black box MPCS. The result is a Reference Model for MPCS Management. Sections III and IV of Figure 3.4 illustrate this decomposition step. Part III of this work describes the decomposition process.

The Reference Model for MPCSs and MPCS Management describe the global tasks for the MPCS and Management components. As a next step, we can completely specify the functional behaviour of the components.

3.3.3 Further Steps

The Reference Model for MPCS Management and MPCS Execution define a structure of components with globally defined tasks. These task definitions can be used as a basis for the development of complete functional specifications of these components, as described in Section 3.2.3.

Section IV and V in Figure 3.4 illustrate the development of the specifications. These specifications can be used as a starting point for a physical implementation, illustrated by the transition from Section V to VI in Figure 3.4.

3.4 Decomposition Strategy

Since decomposition plays such an important role in the development of an MPCS, we will elaborate on a strategy to decompose systems. It should be noted that the definition of 'decomposition' allows an infinite number of decompositions: each decomposition is valid as long as the resulting structure of system components realises the global task of the system that is decomposed. However, we mentioned earlier that we look for particular decompositions, i.e. those that render a structure with components that execute relatively simple tasks. More specifically, we look for decompositions that render a system with a 'natural' structure and division of tasks, which has the conceptual integrity and unity of view point that allows a human to master a complex system as a whole.

We propose to decompose an MPCS by:

1. analysing the requirements that the environment imposes on the system to be decomposed so as to identify which tasks and MPCS has to execute; and

2. allocating the tasks to components so that a natural organisation is created.

We will now discuss and motivate this strategy.

Analysis of Interactions. We analyse the interactions of a component and its environment, and question: "what are the essential characteristics of the interactions and, therefore, can we deduce an organisation of the components from the characteristics of the interactions?".

> **Example.** An MPCS receives forecasts of production targets and firm production targets. The firm targets are due within one week. However, it takes the MPCS one month to transform raw materials into these products. Suppose the MPCS is able to dispatch the products despite the short notice of the firm production targets. We may conclude that the MPCS reacts to forecast targets, and starts manufacturing the products on the basis of forecast targets so that it can quickly dispatch the products when it receives the firm production targets.

We can apply a similar analysis to each component that we have identified as a result of a decomposition step. The strategy can lead to a step wise decomposition since these components may also be eligible for decomposition.

Since we pursue a flexible MPCS, and since the analysis of interactions is a basis for its development, a fairly complete view of all possible situations in which the MPCS and its components may need to interact is essential.

Natural Organisation. References [26,33] describe a few abstract, aesthetic principles for designing systems. They motivate them with the requirement that a human should be able to master and use a system, even when it has considerable, inherent complexity. This applies equally to the designer of the system. The better a system meets this requirement, the better its economies in terms of costs of design, learning and use.

As the principle criterion to meet this requirement, we suggest that the system should have the conceptual integrity and unity of viewpoint that allows a human to master a system as a whole [26]. These characteristics are believed to underlie all principles of quality of system design. It tells the designer not to link what is independent, not to introduce what is immaterial and not to restrict what is inherent. This leads us to three maxims, i.e. 'separation of concerns', 'propriety', and 'generality'.

1. Separation of concerns.

A decomposition of a system should result into components that have essentially different or relatively independent tasks. We call this 'separation of concerns'.

Relatively Independent Tasks.   Separation of concerns leads to relatively independent components that can be viewed as distinct entities in the context of a larger system and therefore can be understood more easily.

Separation of concerns also leads to efficient co-operations of components because the more components can execute their task independently, the less they have to interact. A component's state changes more often as a result of internal, local state transitions than as a result of interactions with its environment. Separation of concerns will, therefore, be reflected in the parsimony of interactions of components.

Separation of concerns can also minimise the effect of a modification of an individual component, on other components [84]. This enhances the generality of an MPCS because changes can be kept locally. Separation of concerns exhibits the inherent parallelism or loose coupling of the system to be decomposed.

Example. The concern of determining that something should be done is delegated to one component and the concern of actually doing that to another. The two components can limit their interactions to commands that describe what should be done and reports about the performance and conditions with which it will be done, is being done, or has been done.

Concerns which are not essentially different but which are nevertheless independent should also be delegated to different components, because it leads to tasks that can be considered in isolation.

We should not separate essentially similar or insufficiently independent concerns. That would increase the number of components and interactions without essentially decreasing the complexity. 'More of the same' unlike 'essentially different' thus does not really contribute to complexity and should not lead to separation of concerns. Consider, for example, the process of selecting a book in a library. The essential mechanism for finding a book, via catalogues, indices and alphabets is similar in a small and a large library [96]. Note that if we decomposed for reasons of performance rather than to achieve a consistent design, we might decide to delegate similar tasks to different components that can execute them concurrently.

Nearly Decomposable Systems. The interactions, although relatively infrequent, are not negligible; otherwise the system would be decomposable in completely independent components. It is somewhat paradoxical that it is precisely in these infrequent interactions the essential raison d'être of a complex structure resides [38].

Sub Optimisation. As a result of separation of concerns, distinct components process different kinds of information. A component therefore has access to only a limited amount of information. The more information a component could use, the better the quality of its decisions could be, assuming that the component has the capacity to digest all information. Separation of concerns therefore implies sub optimisation. However, we have argued in Section 3.1 that an MPCS cannot be optimal in the traditional sense of performance, and that we should strive for satisfactory performance.

However, the decomposition should not deprive components of the essential information they need to do their task. Proposing a decomposition is therefore intimately linked with assessing the relevance of information for a certain task. It should be assessed whether information is relevant in the vast majority of envisaged situations, as we will now explain.

Suppose that one wants to write a letter to a person at a certain address. One posts the letter, and one is not concerned about the path the postman takes to bring the letter to the addressee. This is an example of separation of concerns: the concern of determining the contents of the messages and selecting the addressee is separated from the concern to transport the letter. But, what if the writer of the letter was aware of a likely event that would block all the roads to the addressee and hence prevent the postman from reaching the addressee? The writer might decide to send the letter to someone else, assuming that writing to someone else can have the same effect as originally envisaged. We have therefore given a counter example where it would be better if the concern to write and select an addressee is not separated from the concern to transport the letter. However, the probability that the situation described in this example takes place is quite low. We may conclude that the proposal to separate the writing and transportation is a workable proposal and does not lead to a degradation of the overall system, except for a few, unlikely situations.

We will therefore apply separation of concerns on the basis of a probabilistic analysis to ensure that the likelihood of a performance degradation of the MPCS as a result of the applied decomposition is low. The development of a Reference Model has a heuristic character [4].

Reference [102] gives extensive motivation to apply separation of concerns.

2. Generality.

   If we have identified separate tasks, we should define them in general terms. This leads to components that can execute their tasks for multiple purposes. It enhances the general applicability of the components and hence potentials for future use with unforeseen requirements.

3. Propriety.

   If we have identified separate tasks and defined them in general terms, we should not extend these tasks with responsibilities that are not essential for the whole system. Only tasks that are proper to the essential requirements of the whole system should be considered. Responsibilities that are not proper to the task of the system as a whole would distract us from the proper function and hence conflict with our requirement of consistency. They could also lead to functions that are provided in alternative ways, forcing redundant knowledge and unnecessary choice upon the user.

   Propriety aids open-endedness: by not prescribing about partly-understood needs, one avoids foreclosing the future [26].

The maxims are abstract and qualitative. Proposals for a Reference Model will therefore primarily be assessed in a qualitative sense. As is always the case with lists of aesthetic principles, some are conflicting with others; designing a system is a matter of balance, judgment, and taste.

3.5 Reference Models as a Basis for Formal Specification

In Section 3.2.3, we explained that a Reference Model of a system can be extended so that complete functional specifications of the behaviour of the system's components result. Each specification should define a component that interacts with its environment

---

[4]Reference [67] also considers the applicability of heuristics in a probabilistic sense.

to realise its global task defined by the Reference Model. Moreover, all the components together should fulfill the global task of the entire system, as defined by the Reference Model.

Role of Formal Specification Languages. As designers in general, and as designers of an MPCS in particular, we need to develop precise specifications. Only precise specifications leave little or no room for ambiguous interpretations. Ambiguous interpretations could lead to components that cannot interact usefully.

Formal specification languages are important tools to develop precise specifications because these languages, as opposed to natural languages, have a precisely defined syntax and semantics.

Other benefits of the use of formal design techniques are:

- readable, understandable, flexible, and maintainable descriptions and a framework for their analysis, if the technique offers means to structure the description of a specification and means to develop concise descriptions; and

- formal analysis to proof equivalence of behavioural properties of various descriptions, or to proof the refinement and decomposition of a design, or to validate or test implementations, if the technique is based on a suitable mathematical model.

It would be desirable to express designs of MPCSs in a suitable formal language. However, we know of no formal techniques as of yet that allows us to formally describe a Reference Model.

Informal Reference Model. It is an essential characteristic of a Reference Model that it defines the tasks of its components globally, without completely defining their behaviour. The model would loose its generality, and therefore much of its usefulness as a starting point for the development of specifications for various applications of an MPCS, if it were to specify tasks of components completely.

A complete definition could require, for example, that a particular MPCS component has to schedule the activities of so and so many other components, that these activities would require so and so much time, that the scheduling objective is to optimise throughput, lead times, etc. However, this specific scheduling task would only be useful for an MPCS in some specific application.

The Reference Model therefore defines the scheduling task more globally by characterising the information required to make a scheduling decision, the information that is generated as a result of the scheduling decisions, the activities that are to be scheduled, etc.

A Reference Model specifies global tasks rather than concrete tasks. Consequently, it:

- is a generally applicable starting point for the development of complete functional specifications of components of the complex MPCS; and

- cannot be formally specified with known techniques.

The mapping of an informal Reference Model on complete functional specifications of MPCS components will be informal as well. This should not be interpreted as a denial of the importance of the Reference Model. The Reference Model is extremely important because it reveals a structure and relatively simple tasks of an MPCS that allows us to understand it and to specify it completely despite its formidable complexity. The informal, conceptual characteristics of an MPCS have to be developed and formulated before an attempt to formalise the design or control of an MPCS is appropriate. A Reference Model is the result of conceptual design steps, preceding a complete functional design of an MPCS.

In contrast to the above, it is possible to formally specify the complete functional behaviour of MPCS components.

Formal Specification of Complete Functional Behaviour.   In fact, we do have extensive experience in the formal and informal specification of MPCS components on the basis of the Reference Model developed in this work.

We have developed specifications for Workstation Controllers [15,20]. These specifications have been implemented [107,106] to produce a prototype Workstation Controller for experimentation in a laboratory [18], and its essential control mechanisms and data structures can be recognised in an industrial product [25].

Furthermore, we have developed specifications for transport systems [12,28,27] and Cell/Line Controllers [21].

Our development of the above specifications has given us important feedback to define the concepts of the Reference Model in more rigorous or more generalised terms.

Most of these specifications have been expressed in 'LOTOS', a language developed to formally specify systems with interacting components [31,32,100,65,103]. Reference [17] reviews the use of LOTOS for the specification of MPCS components and gives some simple example specifications and formal analysis of those specifications.

LOTOS has been developed for the formal specification of the complex protocols and services in computer networks. However, its generic concept of interaction, its temporal ordering principles, and its mechanisms for process abstraction are also useful for the formal specification of MPCS components [17]. Moreover, although the interactions in the computer network applications serve to establish data values, the language also allows the definition of types as products and operations and the specification of interactions where products are exchanged or production targets are negotiated, or interactions between devices and materials.

Observational Equivalence. Our [5] definition of 'decomposition' in Section 3.2.1 implies that observers external to an MPCS cannot distinguish between the black box MPCS and the decomposed MPCS as defined by the Reference Model for MPCSs.

The same should apply to the complete functional specifications of MPCS components that are to be developed on the basis of the Reference Model. Consequently, it should not be possible for the environment of the MPCs to distinguish between a system that behaves according to a specification that describes the complete functional behaviour of:

- an MPCS as a black box; and

- a parallel composition of all interacting MPCS components.

In other words, we require that the specifications of MPCS components must adhere to some 'observational equivalence relation' [6].

We can formally express and validate observational equivalence relations in a specification language like LOTOS. Suppose that, according to the Reference Model, an MPCS is decomposed into two components, i.e. a 'Factory Controller' and 'Cell/Lines'. The Factory Controller interacts with the Company Controller and the Cell/Lines. The Cell/Lines interact with the Customers, Suppliers, and the Factory Controller.

---

[5] The remainder of Section 3.5 can be skipped by people who want to focus on the development of Reference Models rather than on formal aspects of complete functional specifications.

[6] This informal definition suffices for this text. Readers interested in the subtle aspects and mathematical properties of equivalence relations of specifications are referred to [65,78].

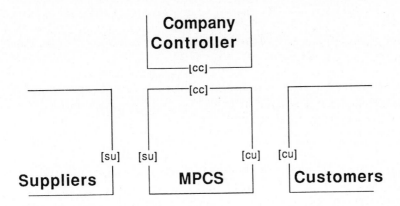

Figure 3.5: A Process Representation Of An MPCS

Let,

process *MPCS*[cc,cu,su] : **noexit** :=
        (* specification of complete functional behaviour of an MPCS
        as a black box. *)
endproc

be a LOTOS-like specification of an MPCS as a process that interacts at gate:

- 'cc' with the Company Controller to exchange production targets, reports material requirements, and reports about the actual production.

- 'su' with Suppliers to exchange raw materials; and

- 'cu' with Customers to exchange products.

Figure 3.5 depicts an MPCS as such a process. Figure 3.6 depicts an MPCS as a parallel composition of processes '*FactoryController*' and '*Cell/Lines*'.

Let,

process *FactoryController*[cc,int] : **noexit** :=
        (* specification of complete functional behaviour of Factory Controller. *)
endproc

50

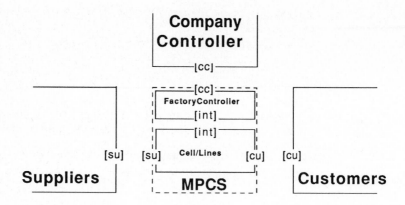

Figure 3.6: A Process Representation Of Factory Controller And Cell/Lines

be a specification of a Factory Controller as a process interacting with Cell/Lines at gate 'int', and let

process *Cell/Lines*[cu,su,int] : **noexit** :=
        (* specification of complete functional behaviour of Cell/Lines. *)
endproc

be a specification of Cell/Lines as a process interacting at gate 'int' with a Factory Controller, at gate 'cu' with Customers, and at gate 'su' with Suppliers.

    Further, let '≈'
denote an observational equivalence relation such as the observation congruence relation in [65].

We then require that a parallel composition of *FactoryController* and *Cell/Lines*, whereby *FactoryController* and *Cell/Lines* can interact at 'hidden' gate int whereas the environment cannot interact at that gate, is observational equivalent to *MPCS*:

MPCS[cc,cu,su] ≈ **hide** int **in** (*FactoryController*[cc,int] |[int]| *Cell/Lines*[cu,su,int]).

    Summarising the above discussion, we conclude that we are not able to formally specify a Reference Model since we know of no formal techniques that are suited to describe the essential characteristics of a Reference Model. Hence, we are forced to describe the Reference Model in a natural language in such a way that the descriptions help the reader to form a mental image of an MPCS as a structure of cooperating components with globally defined tasks.

However, our own experience has shown that a Reference Model can be used as a basis for the formal definition of complete functional specifications of MPCS components so that these components can be subject to formal analysis.

We will include some simple but formal pointers to the development of complete functional specifications on the basis of the Reference Model. In particular, we will point to formal decompositions and equivalence relations as described above.

3.6   Focus on Reference Model for MPCSs and MPCS Management

In this book, we will focus on the development of a Reference Model for MPCSs and MPCS Management, being a necessary first step to design an integrated and flexible MPCS, by defining a structure and global task definitions for their components.

These Reference Models are generally applicable by their very definition, but nevertheless have far-reaching consequences as early specifications in the top-down design of an MPCS. The strategic importance of Reference Models to achieve the integrated automation discussed in Section 1.1, its potentially wide-spread use, and the unavailability of reported literature about a Reference Model for MPCSs and MPCS Management is a motivation for developing these Reference Models.

In addition to being a baseline in the development of integrated and flexible MPCSs, the Reference Models will also serve as a source of reference in discussions about MPCSs.

Source of Reference.   People can discuss aspects of components such as required behaviour, implementation, costs, and standardisation meaningfully only if they have a common understanding of the functionality of the component, and how it fits into the overall MPCS scheme. This information is contained in the Reference Model.

A common nomenclature would be very welcome. Today, even the words that are most frequently used by the manufacturing community have many interpretations. 'Master Production Schedule', or 'MPS', for example, is used to denote the:

- production targets for a factory; or

- input to an MRP [105] system; or

- dispatch schedule made by a factory in a make-to-stock situation.

People, who know the framework of the Reference Model developed in this work, will recognise that these three interpretations are all different.

Development of a Reference Model.   Several papers discuss the need for a Reference Model for MPCSs [14,37,62,81,110]. References [47,77] describe organisations in general, but do not discuss MPCSs as such. Nevertheless, some of their general conclusions have been applied to the development of the Reference Model that is described here. References [2,12,13,29,34,61,68,73,74,105] describe MPCSs, but only parts of it. They focus on such tasks as production planning, transport systems, or robot control, or sensing.

References [3,4,8,63,14,36,40,66] describe an MPCS in a more complete sense. However, these models are imprecise, ambiguous, and incomplete to serve as a basis for a further design of an MPCS. In fact, some of them do not define the tasks of MPCS components at all. Others do, however without any justification for the definition of tasks and structure. Reference [85] reviews several Reference Models.

The model developed in this work differs from all of the models mentioned above: it encompasses the complete MPCS and its Management and it is justified on the basis of criteria that support the development of a consistent, conceptually uniform model, which can be easily understood [7]. This will lead to significantly different results.

Complete Functional Specifications.   We will not develop complete functional behaviour of MPCS components. However, we will give pointers to these specifications to illustrate the global task definitions of the Reference Model. When we discuss the specifications, we mainly discuss interaction primitives and refrain from defining their temporal ordering. We will also point to the observational equivalence relations discussed before.

You are referred to references [15,20,28,27] for some of the complete specifications we have developed, and to references [17,90,99,103] for discussions relevant for making formal specification of MPCS components.

---

[7]References [22,23] can be considered as reviews of the model developed here, although they differ in some respects.

Part II

Reference Model for MPCSs

Chapter 4

Decomposition of an MPCS

In chapter 2, we described the interactions of an MPCS and its environment. In this chapter, we follow the strategy described in Section 3.4, to analyse these interactions in order to identify relatively independent tasks of an MPCS. Subsequently, we decompose the MPCS and allocate these tasks to MPCS components.

Recall that, in Part II of this work, the MPCS is assumed to have a stable product portfolio, production capacity, and production costs.

4.1 Analysis of MPCS Interactions

Realising a Required Service Level. An MPCS receives firm production targets from a Company Controller, instructing it to dispatch specific products at certain due dates. It also receives *forecast* production targets from the Company Controller, indicating which products may be called upon at certain due dates. The MPCS informs the Company Controller which raw materials it requires and when, so that it can realise the production targets.

Let us first consider the following definitions relating to production targets.

A 'production target' is a request to an MPCS to accept certain raw materials from certain Suppliers in certain time slots, and to dispatch certain products to certain Customers in certain time slots.

The 'service level' is a measure for the extent to which an MPCS can meet production targets.

56

The 'order quantity' is the number of units of products to be dispatched according to a production target.

The 'order cycle time' is the time between the moment an MPCS receives a firm production target for a specific product and the moment the products are due.

An MPCS is capacitated: it can manufacture a limited number of products, needs a certain time to manufacture them, and needs time to switch to the manufacturing of other products. The Company Controller may therefore propose production targets that are infeasible for the MPCS with respect to the required quantities or due dates. However, we will analyse how an MPCS can increase its responsiveness to production targets with varying order cycle times and order quantities by exploiting forecast production targets.

Varying Order Cycle Times.   Order cycle times may be long enough, for example, to allow an MPCS to obtain raw materials after it has received production targets, and transform these into the required products in time. However, at another point in time, the order cycle time of a particular production target may be shorter than the time the MPCS would need to obtain raw materials and to transform these into the required products.

An MPCS can only respond to the production targets with the short order cycle times quickly enough by taking partially processed materials out of stock and transforming them into the required products. Partially processed materials are raw materials that have been processed to some extent so that they need less time to be transformed into products than raw materials would. The MPCS should create these partially processed materials, and replenish them, on the basis of forecast production targets. When the order cycle times are short, it deletes its inventory; it has a chance to replenish them when they are long.

Figure 4.1 illustrates the use of inventory to accommodate varying order cycle times.

An order cycle time may vary, but order quantities may vary as well. Take, for example, products with seasonal demand patterns.

Varying Order Quantities.   Typically, an MPCS cannot change its capacity quickly and economically. Varying order quantities therefore tend to lead to periods in which the MPCS does not fully utilise its production capacity and other periods in which the MPCS does not have enough capacity to supply the required products.

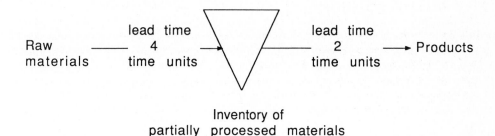

Inventory of
partially processed materials

Suppose that, given the current load of an MPCS, it takes six time units to transform raw materials into products, four units to transform raw materials into certain partially processed materials, and two units to transform these into the products.

If the order cycle time is three units, production targets can be realised within the order cycle time if the MPCS contains sufficient inventory of the partially processed materials, which can be transformed within two units. This inventory needs to be created on the basis of forecast production targets, given by the Company Controller.

If the order cycle time exceeds six units, production targets can be realised by processing raw materials, assuming that those are available.

To achieve a high service level for production targets with short order cycles, the MPCS must ensure that there is a high probability that it contains sufficient inventory to realise the production targets.

Figure 4.1: The Use Of Inventory To Accommodate Varying Order Cycle Times

Under utilisation implies inefficient use of resources, whereas inability to supply the required products degrades the service level.

However, an MPCS may be able to better utilise its capacity and also be able to respond to production targets with order quantities that exceed its capacity. It should process raw materials and keep partially processed materials in stock when order quantities of firm production targets are small whereas order quantities of forecast production targets exceed its capacity. It therefore utilises its capacity in times of small order quantities by producing inventory and can respond to large order quantities by taking partially processed materials out of the inventory to transform these into products.

We conclude that an MPCS can improve its responsiveness to production targets with varying order cycle times and order quantities despite its limited capacity by exploiting forecasts to create inventories. However, as we will explain, inventories have some undesirable properties so that an MPCS should limit its inventories.

Negative Effects of Inventory.   Inventories incur operational expenses, claim space, may become obsolete, increase the manufacturing lead time, may delay feed back about quality, etc. Moreover, inventories can also have a negative impact on the service level of an MPCS. The reason is that an MPCS uses a portion of its production capacity to create inventories –production capacity that it might have been used instead to produce inventories or products that are needed more urgently.

> **Example.** As Figure 4.2 illustrates, an MPCS can transform materials A into materials B, and materials B into products C. It can therefore produce C by processing A or B, but needs less time if it takes B.
>
> The MPCS has to dispatch C. It has already transformed many A's into B. As a result, it has enough of B to produce C, but no longer enough of A to produce C. It will therefore take B and produce C.
>
> However, the order cycle time is much longer than the time it needs to process B, whereas the order cycle time would be just long enough to process A. The MPCS has therefore produced B well before it actually needs them. It has therefore used a portion of its capacity which could have been used for more urgent production instead of producing B.
>
> It may therefore have been better if the MPCS had not yet processed A into B so that it had a sufficient amount of A in stock rather than B.

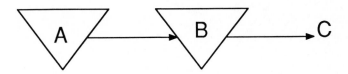

Figure 4.2: Products C, Resulting From Processing Materials A Or B

Example. As Figure 4.3 illustrates, an MPCS can dispatch products E
and F. Both products have a common material D. The MPCS has to dispatch products E, but it will not be able to dispatch them in time because
it does not have enough D. The reason is that it has already processed D
to make products F. It could therefore have dispatched product F if these
were required.

The MPCS might have been able to produce the required products E if it
still had D in stock. It would have been better if the MPCS had not yet
processed D and produced products F so that it could processes them now
to make products E.

A subtle interplay exists between inventory and production capacity. We concluded
earlier that inventories can be used to increase the service level. The examples, however,
illustrate that inventories may also have a negative impact on the service level. An
MPCS is therefore faced with a problem: which inventories to produce, and when, so
as to improve the service level while limiting the costs associated with inventories.

Selection of Materials and Timing of their Production. The above examples also illustrate that an MPCS does not necessarily have to produce inventory
of products or nearly-finished products to improve its service level. It can process raw
materials a bit and postpone further processing till the forecast production targets
become more firm; it maintains inventories of several partially processed materials,
as Figure 4.4 illustrates. Indeed, it can improve its service level and the utilisation

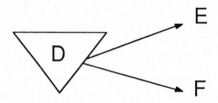

Figure 4.3: Materials D, Which Can Be Transformed Into E Or F

of its production capacity by planning the manufacturing. Raw materials and partially processed materials can be kept in the 'upstream' stocks and, while production targets become more firm, be transformed into materials of the 'downstream' stocks. In the upstream stocks, the materials have consumed little capacity, can possibly be transformed into several products, but need some time to be transformed into products; in the downstream stocks, the materials have consumed more capacity, can be transformed into fewer products, but can be transformed into products more quickly.

A 'multi-stage' or 'multi-echelon' production system' is an MPCS that distinguishes several production stages decoupled by inventories of raw materials, partially processed materials, or products, and that coordinates the manufacturing at these stages to realise a required service level while limiting the resulting costs.

An 'item' is a raw material, partially processed material, or product at the beginning or end of a production stage in a multi-stage production system.

> **Example.** An MPCS manufactures several mainframe computers. Customers can choose from a variety of products with different memory boards, I/O boards, disk drives, processors, etc. The MPCS manufactures items such as memory boards and disk drives on the basis of forecast production targets. This is one stage of the production. In the next stage, it takes these items to assemble the specific computers when it receives firm production targets for these computers.

We conclude that an MPCS may use a multi-stage production structure to increase its service level despite its limited capacity.

Multiple Production Stages. So far, we have described the use of inventories that separate the various production stages, as a protection of a capacitated MPCS against uncertainty in the timing and volume of demand. Similarly, the inventories can be used as a protection against uncertainty in the supply of raw materials and in the processing of items.

> **Example.** An MPCS manufactures TV sets. In one stage, it keeps an inventory of raw materials like Integrated Circuits. Because of this inventory, the supply of raw materials to the other production stages is reasonably reliable despite the unreliable supply of raw materials by suppliers. In other stages, the MPCS assembles the TV sets as outlined by production targets.

> **Example.** An MPCS produces cartons of pre-cooked, deep-frozen spinach. In one stage, it keeps an inventory of fresh spinach. It takes in enough spinach during the season to accommodate a year's demand. In another stage, it processes the spinach, all year round.

An MPCS may also have several production stages separated by inventories to achieve economies of scale. It may be prohibitively expensive to produce exactly the number of items required in a response to a certain production target. The reason is that suspending the manufacturing of one item and switching to the manufacturing of another item requires change-overs of machines, during which the machines are not productive. In such cases, it may be more economical to reduce the change-overs and make the items in batches of a size larger than the required number, which will result in a temporary inventory of items.

> **Example.** An MPCS manufactures TV sets. In one stage, it produces Printed Circuit Boards for all types of TVs. It produces them in large batches to sufficiently utilise the expensive component placement machines. Hence, an inventory of PCBs will be generated. The PCB assembly stage supplies PCBs from its inventory to other stages that make certain types of TVs.

We will now analyse how multiple production stages should be coordinated, i.e. when the various production stages should produce the items.

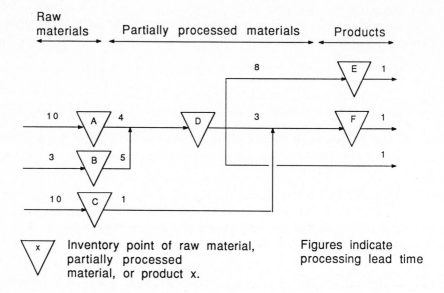

Figures indicate processing lead time

Inventory point of raw material, partially processed material, or product x.

The order cycle time is two time units, and an accurate forecast of the production targets exists six time units in advance. Products E and F can be dispatched within the order cycle time. Processing of D into F can still be started on the basis of accurate forecasts between six and four time units before F will probably have to be dispatched. However, transformation of D into E cannot be based on an accurate forecasts because the processing time exceeds the six time units. It has to be based on less accurate forecasts. The required stock of D depends on the accurate forecasts of F and the less accurate forecasts of E, but also on independent demand for D. The required stock of D in turn creates a demand for raw materials A and B.

Figure 4.4: Multiple Production Stages

Multi-Stage Production Coordination.   We need some more definitions to elaborate on multi-stage production coordination.

An 'inventory point' is an inventory of a specific type of items in a multi-stage production system.

'Work-In-Progress', or 'WIP' are items that are released from an inventory point to be transformed into items of another inventory point and that are still being processed.

An 'inventory position' is the type of items contained by an inventory point.

An 'inventory level' is the quantity of items contained by an inventory point.

A 'job' is a transformation of items from inventory points into items of other inventory points, the addition of items to, or removal from inventory points.

The 'job lead time' is the time needed to execute a job.

The 'job cycle time' is the time between the moment it has been determined that a job should be executed and the moment that the job is due.

The 'commonality' of an item is a measure for the extent in which the item can be used for the manufacturing of multiple product types.

Information Required for Multi-Stage Production Coordination.   Contrary to single-stage production systems, the time to replenish an inventory point depends not only on the job lead time to produce the replenishing items but also on the availability of items in the 'upstream' inventory points [35]. An overall view of all inventory levels and job lead times may therefore be required to determine which jobs should be executed.

Here are some data that should be taken into consideration when determining when a certain job should be executed:

- the firm and forecast production targets;

- the availability, capacity, and capability of resources to execute the jobs;

- the availability of items; and

- the costs, job lead times, and precedence relations of jobs.

Some of these data tend to vary slowly during the time an MPCS has to realise a production target. This applies to:

- the precedence relations of job, because they depend on the way the products are designed and assembled; and

- capacity and capability of resources because they depend on their physical implementations.

These data can be viewed as given constraints when determining which jobs to execute because they change so slowly.

However, the MPCS cannot assume that the job lead times have a given value. They depend on the utilisation of the MPCS production capacity and therefore on the production targets and the MPCS's decision when to execute which jobs. Similarly, the ordering, processing, and holding of items incurs costs which may depend on the utilisation of an MPCS's production capacity: executing a job when the MPCS is under utilised may be cheaper than, for example, when it is fully utilised and would imply the disruption of on-going jobs. The actual costs should therefore be known when determining when, which jobs should be executed.

The availability of items is another set of data that depends on the actual execution of jobs. Since jobs may be late or may produce defect items, it is important that the actual availability of items is known when determining which jobs to execute.

We conclude that information about forecast and firm production targets, actual and expected job lead times, availability of items, and various costs of executing jobs may be required to coordinate multiple production stages. However, it is not important to know how the jobs are executed to decide whether and when they should be executed. The decision, for example, to execute a job that transforms a Printed Circuit Board and other items into a TV in a certain processing lead time does not depend on the fact that the PCB is subject to component insertion, soldering, inspection, and repair operations. The task to determine the required inventory levels and to determine when and which jobs should be executed can therefore be executed on the basis of knowledge of production targets, job lead times, availability of items, and the various costs of jobs. This task can be executed, relatively independently from the task of executing the jobs.

Independent MPCS tasks.   To summarise, we have identified the following, relatively independent, globally defined tasks of an MPCS:

- **Production Planning**: realise a certain service level while keeping inventory costs low by determining the required levels of inventories that decouple the production stages and by determining which jobs should be executed and in which time slots to realise the required inventory levels; and

---

Production Planning

Knowledge:

- how to react to stochastic events as demand and supply;
- how to plan the use of production capacity; and
- of lead times, yields, and costs of processing and holding items.

Execution of jobs

Knowledge to:

- process items reliably and cost-effectively; and
- assess whether jobs can be executed in certain time slots.

Figure 4.5: Knowledge Required For MPCS Tasks

- **Execution of Jobs**: determine whether certain jobs can be started and completed in certain time slots and execute these jobs when requested to do so.

These tasks are also quite different. The production planning task serves to ensure that the right items are available so that forecast production targets can be realised. The task to execute jobs deals with the transformation of items that are available. The production planning task is heavily affected by the uncertainty of production targets. It implies planning and decision-making with uncertainties –reacting to firm and forecast orders while weighing the risks of late deliveries or inventories. The task to execute jobs has to deal with firm orders only. It deals with a deterministic demand because the decisions to execute jobs have been taken. Figure 4.5 illustrates the types of knowledge and abilities required to execute the MPCS tasks.

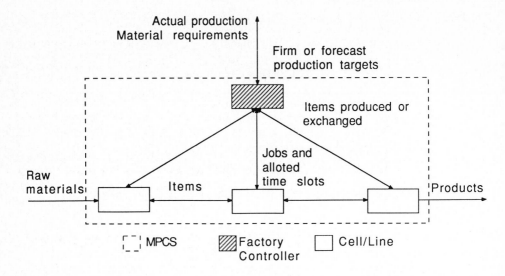

Figure 4.6: An MPCS Decomposed Into A Factory Controller And Cell/Lines

4.2 Factory Controller and Cell/Lines

In compliance with the development strategy discussed in Section 3.4, we assign the independent MPCS tasks, identified in Section 4.1, to distinct components, i.e. a 'Factory Controller', which has to coordinate production stages, and a 'Cell/Line', which has to execute jobs. A Factory Controller determines which job a Cell/Line should execute in which time slot. A Cell/Line reports which items it has produced or exchanged. Figure 4.6 illustrates our proposal for an organisation of an MPCS.

One Factory Controller.   We concluded that the required inventory levels depend on the demand for all inventory points. The overview of all inventory points and their mutual dependencies can best be obtained by a single Factory Controller. Hence, we propose that the MPCSs have only one Factory Controller.

Multiple Cell/Lines.   Different jobs may require the same resources. Two jobs, each producing a different PCB, for example, may require the same component insertion machine. A Cell/Line, considering a commitment to execute a new job in a certain time slot, must be able to assess the effects of this commitment on earlier commitments since accepting a new commitment may prevent it from meeting earlier

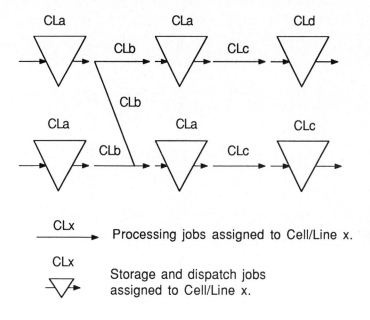

CLx → Processing jobs assigned to Cell/Line x.

CLx Storage and dispatch jobs assigned to Cell/Line x.

Figure 4.7: A Possible Assignment Of Jobs To Cell/Lines

commitments. Jobs that can effect each other's lead time must therefore be assigned to the same Cell/Line. We therefore propose that there may be multiple Cell/Lines, provided that groups of dependent jobs are assigned to the same Cell/Line.

Cell/Lines may have to transform, store, and dispatch items. Typically, the storage and transformation of items require different resources so that they are independent and can be assigned to different Cell/Lines. Often, the storage of multiple stocks will be assigned to a single Cell/Line, containing the central warehouse, which we find in many factories. Figure 4.7 illustrates a possible assignment of jobs to Cell/Lines. In Chapter 11, we elaborate in detail on the assignment of jobs to Cell/Lines.

Observational Equivalence. In Section 3.2.3, we discussed the development of complete functional specifications for MPCS components on the basis of a definition of their global tasks. We have just identified the global tasks of Factory Controller and Cell/Line, as a result of a first decomposition step to develop a Reference Model for MPCSs.

We require that complete functional specifications of Factory Controllers and Cell-/Lines to be developed on the basis of the Reference Model satisfy the following relation: a parallel composition of specifications of Factory Controller and Cell/Lines should be observational equivalent to a complete functional specification of an MPCS. We already expressed this requirement, in a more formal manner, in Section 3.5.

In the following Section, we describe some practical situations and explain the role of the Factory Controller and Cell/Lines in these situations.

4.3 Modeling Real-Life Situations

In this Section, we describe how Factory Controllers and Cell/Lines cooperate to realise an MPCS. We describe some general tasks that a Factory Controller may be required to execute, and describe Factory Controller–Cell/Line interactions in typical scenarios. Further, we discuss some specific requirements for the multi-stage inventory control procedures to be executed by a Factory Controller and compare these with reported procedures in the literature.

General Requirements for a Factory Controller. A Factory Controller may have to operate in a variety of situations. It may be required to execute such tasks as:

- Materials requirements planning. The Factory Controller should report to the Company Controller which raw materials it needs in order to realise firm production targets or to prepare for forecast production targets. It should also alert the Company Controller if these materials are not available in time. It may also require materials to compensate for wasted materials.

- Capacity planning. As explained before, the Factory Controller should efficiently utilise the Cell/Line capacity and determine in which time slots jobs be executed to realise production targets.

  The Factory Controller must be guaranteed that jobs will be executed in a certain time slot. It therefore negotiates with the Cell/Lines whether they can commit themselves to certain time slots. A Factory Controller may give Cell/Lines commands to execute jobs in time slots that start immediately or somewhere in the future. Quite often, practical Cell/Lines are not advanced enough to make such commitments. In such cases, the Factory Controller simply assumes that the jobs will be executed within a certain job lead time, although the variance in the lead time may be large. The Cell/Lines periodically report to the Factory Controller which jobs have actually been executed. The Factory Controller uses this information to determine which raw materials need be ordered.

- Materials coordination. The Factory Controller orders Cell/Lines to execute jobs. These could be jobs to accept raw materials from a supplier, to dispatch products to a customer, to store or dispatch items, or to transform items.

The Factory Controller should ensure that a Cell/Line has timely access to the items needed to execute its jobs. The Factory Controller therefore commands other Cell/Lines to provide these items in a certain time window. The Cell/Lines can negotiate among themselves when in the time window they exchange the items and where.

> **Example.** Figure 4.8 illustrates a particular component distribution configuration, which captures a typical material coordination problem. There are 'production Cell/Lines' and 'storage Cell/Lines', connected by four kinds of component flows:
>
> 1. Components that are passed directly from a supplier to the production Cell/Line. The production Cell/Line tells the supplier when the materials are needed. This could be the bulky components that, in order to limit space usage, should be brought in when they can be processed directly.
>
> 2. Components that are stored in a central store and dispatched from there to a production Cell/Line. These could be components that are needed by multiple production Cell/Lines for a specific job. A particular Cell/Line needs, for example, a specific component to produce 2000 PCBs of a certain kind.
>
> 3. Components that are stored in a central store, and moved to a store close to a production line, from where they can quickly replenish parts processed by the production Cell/Line. These could be components used by multiple production Cell/Lines for multiple jobs.
>
> 4. Components that are stored close to a particular production Cell/Line. These could be components that are specific for one production Cell/Line only but that should be dispatched quickly when the production Cell/Line needs them.
>
> The Factory Controller determines the time slots in which the storage and production Cell/Lines exchange the components. Note that a Factory Controller may have to order a storage Cell/Line well in advance to dispatch materials. Quite often, they have a limited capacity of lifts and carriers and have to move components to a location from where they can quickly be passed to a production Cell/Line well in advance of the actual dispatch.

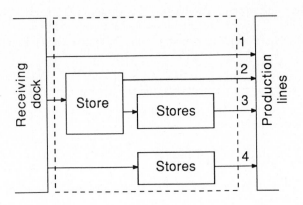

Figure 4.8: Component Distribution Configuration

Let us now review some typical scenarios of Factory Controller–Cell/Line cooperation.

Exchange of Allocated Items. Some items are stored and transported as bulk items in bins, on tapes, etc. It may be too expensive if Cell/Lines pick items out of these bins to exchange precisely the number of items requested by the Factory Controller. Rather, they exchange one or more bins with at least the required number of items and return the unused items while keeping the Factory Controller informed about the exchange of items. The items that are not intended for use are called 'non-allocated inventory'. The items that are intended for use are called 'allocated inventory'. The Cell/Lines should report to the Factory Controller which items they exchange and whether they are allocated or non-allocated.

> **Example.** A Factory Controller orders Cell/Line A to dispatch 500 ICs to Cell/Line B. Cell/Line A has a partly used reel of 30 ICs in stock and unused reels with 1000 ICs. It dispatches 500 allocated ICs and 530 non-allocated ICs from reels of 30 and 1000 to Cell/Line B, and expects the 530 non-allocated, on the reel of 1000, to be returned later.

Overlapping Time Slots. The Factory Controller should coordinate the exchange of items by Cell/Lines. It should do this in such a way that the Cell/Lines can operate as independently as possible despite the fact that they exchange items.

The Factory Controller should therefore refrain from commanding:

- Cell/Line A to accept certain items, process them, and dispatch the processed items, and do all this in time slot X; and

- Cell/Line B to accept these processed items, process them in turn, and dispatch the processed items, and do all this in time slot Y,

if time slots X and Y overlap. The reason is that Cell/Lines A and B would not be able to operate independently since B is dependent on the exact time of delivery of items by Cell/Line A.

In fact, it two Cell/Lines are to exchange items, one of them has to be commanded either to dispatch items that it has in stock, or to accept items and store these. The Factory Controller can, for example, request:

- Cell/Line A to be prepared to take certain items out of stock and to pass these in a time slot X to Cell/Line B; and

- Cell/Line B to accept and process these items in time slot X.

In this way, the items of inventory points are available during the entire time slot for the Cell/Line that has to process the items. This Cell/Line can operate independently.

A Factory Controller should request a Cell/Line to be prepared to dispatch items in a certain time slot well in advance to allow the Cell/Line to pick up the items and move them to the location where they have to be exchanged.

Special Items.   A Factory Controller may have to allocate items to Cell/Lines that are not directly related to jobs. Lubricants and bulk items, for example, cannot easily be allocated to one job, and can better be replenished periodically.

Delivery from Several Inventory Points.   Notice that multi-echelon structures as depicted in Figure 4.9 are possible. The Factory Controller may decide to transform items from inventory points as far upstream as possible given a certain order cycle time. These items are directly transformed into products, bypassing the downstream inventory points.

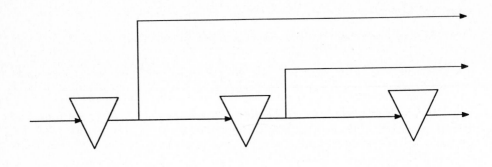

Figure 4.9: Output From Several Inventory Points

Commonality of Items. We have already discussed an example with items with multiple uses. These items are 'common' to several products. The demand for common items has reduced variance compared with the variance in demand for the products, at least if the demands for the products are not correlated. The lower the variance in the demand for items, the smaller the safety margin of the inventory level of items that is required to prevent stock outs. It seems therefore that commonality can be exploited to reduce the inventory costs. This is often referred to as 'risk pooling'. However, as References [6,49] show, we only have approximate models for determining the actual benefits. The benefits depend on the specific definition of the service level and may have to be traded against the increased inventory of the specific (non-common) components and against the costs of replacing specific components by common components.

We will now review some procedures to be executed by Factory Controllers.

Material Requirements Planning. Many real-life MPCSs use 'Materials Requirements Planning' [1] systems [105] to determine which raw materials they need and when. These systems consolidate the firm and predicted production targets. Then they calculate which raw materials are needed assuming a fixed relation of product, raw materials, and manufacturing lead time. Finally, they compare the required raw materials with the available, ordered, and committed raw materials to determine which

---

[1]MRP Systems are also used for scheduling, as we will discuss in Chapter 5.

materials have to be brought in from suppliers.

The common use of MRP systems to determine the raw materials requirements can be viewed as an implementation of a Factory Controller. However, this implementation has a rather limited capability when compared to the control problems described in this chapter. To accommodate for variability in order cycle times and order quantities, typical MRP packages can only maintain inventory of (finished) products by means of overplanning or 'hedging' the demand for products [108]. The inventory level of products required to maintain an adequate service level necessarily has to reflect the uncertainty in demand during the entire lead times from procurement of raw materials up to and including the manufacturing of products.

Typically MRP systems do not assume a multi-echelon structure, do not exploit commonality of items, and do not adapt to varying job lead times. They do not optimise the inventory levels with respect to some objective, and make no explicit provision for dealing with uncertainty. Common MRP systems are simple but do not exploit some possibilities that inherently exist in echelon systems to reduce inventory costs while maintaining service levels.

References [55,105] describe MRP systems. Reference [108] elaborates on the use of inventories and MRP systems. References [59,93] describe MRP systems, as well as alternative approaches that exploit available forecasts of demand, take into account limited production capacities, and that aggregate demands for individual items to reduce variance.

Alternative Control Procedures.   After the critical assessment of MRP, it is fair to note that the available literature does not seem to provide alternative, practical, and generally applicable procedures for multi-stage production coordination. Reading a review in Reference [54], for example, one could notice that many of the proposed multi-stage production coordination procedures follow a paradigm that:

- focuses exclusively on the use of inventory points to accommodate variability in production quantities ignoring variability in order cycle times;

- assumes an unlimited capacity of the MPCS;

- does not exploit forecasts of production targets but assume stationary demand processes;

- assumes known and deterministic job lead times, whereas often job lead times are non-deterministic.

Moreover, most procedures are applicable for simple, serial product structures rather than for the more complex, though common, assembly structures, although

References [71,72] describe a procedure that can be applied to rather arbitrary serial, assembly, and disassembly structures, and combinations thereof.

We have described multi-stage production coordination as a strategy to increase the service level of a capacitated MPCS [2]. The issues are the required service level, the planning of capacity, and the total inventory. However, the typical paradigm in literature ignores the capacity constraints and focuses on minimizing total inventory costs assuming that inventory costs per item are known. These assumptions reflect the common view that value is added to items as they are processed so that the inventory costs of downstream items are higher than those of upstream items. However, the added value cannot easily be assessed. As Reference [51] points out, the value of items can hardly be quantified since they have no customers. The available literature therefore seems to optimise rather artificial cost parameters rather than capacity usage and total inventory levels, which can be measured more accurately.

We need algorithms and procedures for a Factory Controller in an operational scenario. One operational scenario would be a Factory Controller which considers to order a Cell/Line to process items contingent on the time the Cell/Line needs to process them. The job lead time is variable because it depends on the load. However, the Cell-/Line may be able to accurately estimate the lead time. The lead time therefore has predictable uncertainty. The Factory Controller and the Cell/Line negotiate the lead times. If the Cell/Line commits to a certain lead time, the Factory Controller may decide to order the products; other wise, the Factory Controller may decide to explore another avenue.

An Application. Reference [112] describes a practical approach for an MPCS in a make-to-order situation that combines MRP systems and multi-stage production coordination to achieve good service levels. In fact, there are two stages, separated by an inventory point, which is called 'decoupling point'. Items of a decoupling point can be transformed into products within the typical order cycle time on the basis of firm production targets. The decoupling point therefore decouples the customer driven production and the forecast driven production [60].

The products in Reference [112] are "tailor made", customer-specific products with many common items. The decoupling point contains these common items. The demand for the items can be forecast more accurately than the demand for products because of the commonality effects discussed before. The demand for the items drives the MRP system rather than the demand for the products. The inventory levels are

---

[2]Multi-echelon system predominantly appear in sales and distribution organisations where different inventory points carry the same products at different locations. Inventory points close to the customers could be replenished from central inventory points as customer demand becomes more firm [95]. Multi-echelon systems can also be used by an MPCS as explained above.

reduced because of the relative accuracy of the demand that drives the MRP system. The approach described in Reference [112], therefore, selects an inventory point, the decoupling stock, that allows the use of MRP systems with relatively low inventory costs and that allows the dispatch of tailor made products within relatively short order cycles.

The Need for Multiple Production Stages. Many real-life MPCSs will improve their product designs and manufacturing technology and control to manufacture products more quickly so that job lead times can be reduced. This reduces the need for multiple production stages. However, other phenomena will increase the need for multiple stages and inventories in between. Company Controllers may, for example, exploit reduced order cycle times to promise customers quicker deliveries. Quicker deliveries enforce a need for shorter order cycle times and hence for multiple production stages. The on-going proliferation of the products portfolios will have the same effect. Companies may also take over some of the manufacturing of raw materials, which is currently done by suppliers. The job lead time for the MPCS will increase and hence the need for inventory points. Moreover, the need for inventory points to prepare for varying order quantities remains.

Selection of Inventory Points. It depends on the situation which and how many different production stages should be identified. Figure 4.10 illustrates a simplified situation, where four major stages are identified to make televisions, i.e:

- one stage receiving and storing raw materials like bare board Printed Circuit Boards and electric components;

- one stage taking raw materials, assembling, and storing PCBs;

- several stages, each taking PCBs and assembling certain types of TVs; and

- packaging, storing, and dispatching TVs.

This organisation will probably be chosen to allow the PCB stage to produce PCBs in large numbers so as to improve the utilisation of the expensive component placement machines. However, the separate stage for PCB production leads to an inventory of PCBs and to increased throughput times from raw materials to product. If these disadvantages outweigh the advantages, an organisation as illustrated in Figure 4.11 may be preferred.

In this organisation, the stages that assemble the TVs also assemble the PCBs. Each stage makes only those PCBs that are required for its TVs. This typically implies that

76

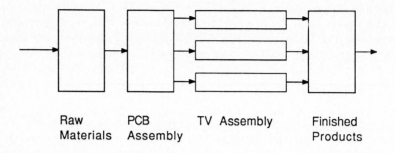

Raw   PCB   TV Assembly   Finished
Materials   Assembly   Products

Figure 4.10: Four Stages For TV Assembly

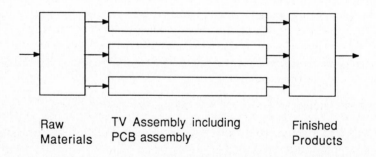

Raw   TV Assembly including   Finished
Materials   PCB assembly   Products

Figure 4.11: Three Stages For TV Assembly

the PCBs are made in small batches and that the component placement machines have to change over often. As a result, the PCB assembly as such will be less efficient than in the structure of Figure 4.10. Also, the structure in Figure 4.10 will need a little more time to dispatch a product after reception of a production target than the structure in Figure 4.11 because it cannot take PCBs out of stock. On the other hand, there will be less inventory and the throughput time from raw material to product will be less.

Depending on the relative costs of inventory and under utilisation of component placement machines, the structure of Figure 4.10 or 4.11 is more preferable.

Volume and Mix Commitment.   Finally, we discuss how MPCSs in a make-to-stock situation often receive their production targets in multiple phases. Often, the Company Controller can more accurately forecast the production targets for aggregate groups of products that share such marketing characteristics as price, styling, and options. The forecast production targets are therefore expressed as volumes for these groups. As forecasts of demand become more accurate, the volume targets are replaced by mix targets, which specify the required mix of products within the groups.

The Factory Controller should agree on the targets for aggregate groups of products. It should therefore assess their feasibility by considering which jobs will have to be executed, when, and by which Cell/Lines to realise them. The Factory Controller can assess the feasibility of production targets for several aggregate groups independently if they do not require the capacity of the same Cell/Lines and do not require the same raw materials or items. In other words, the Factory Controller can most easily assess the feasibility of production targets for an aggregate group of products that exclusively need a critical resource.

Example.   TV sets with and without stereo sound belong to different commercial groups. However, the Factory Controller cannot independently assess the feasibility of production targets for these groups if they contain the same size picture tube, which can only be assembled by one Cell/Line.

As follows from the above, a Company Controller can best forecast the targets for an aggregate group of products that share certain market characteristics, whereas the Factory Controller can best assess the feasibility of production targets for groups of products that share certain manufacturing characteristics. The Company Controller and Factory Controller should therefore negotiate groups of products that share both commercial and manufacturing characteristics.

4.4   Factory Controller–Cell/Line Interactions

We will give some interaction primitives to illustrate a Cell/Line [3]. As explained in Section 2.2, we give interaction primitives to model MPCS components at a high level of abstraction. We give an interaction primitive, for example, to model a Factory Controller and Cell/Line that agree to execute a job. We do not give interaction primitives to model possible negotiations that lead to such an agreement, with, for example, the Factory Controller proposing quantities of items to be produced in certain time slots and the Cell/Line accepting or rejecting the proposals.

Table 4.1 contains interaction primitives that can be used to model a Cell/Line that:

- Agrees with a Factory Controller to execute a job in a certain time slot by means of an 'Operation-Command'. A job defines the items involved in a transformation. The items to be processed may have to come from another Cell/Line or may be already contained in the Cell/Line. The items to be produced may have to be dispatched to another Cell/Line or may have to be stocked by the Cell/Line. The Factory Controller may overrule an Operation-Command by issuing a new command with the same 'Command-Id'. A command that overrules previous commands with the same Command-Id can be used to relax the due date, to change the numbers of products, etc;

- Can give the Factory Controller information about inventory of items for example to report that items have disappeared or cannot be located;

- Can report about the progress of a job. There may be several Progress-Reports referring to the same Operation with a Command-Id. The parameters define the items that have been exchanged, are kept or put in stock, and what kind of progress has been made.

  A 'completed' Progress parameter indicates that the Operation has been executed. The items mentioned describe the result of the execution. These items may differ from those mentioned in the Operation-Command if, for example, defects have been produced or more items have been used than ordered.

  A Progress parameter 'busy' can be used to indicate which items have been exchanged as allocated or non-allocated inventory.

  A Progress parameter 'waiting' indicates that a job cannot be executed because items are not there. The item parameters are used to indicate which items are available, have already been accepted, etc.

---

[3]Appendix A describes the meaning and role of interaction primitives.

Interaction Primitive	Arguments	Explanation
Operation-Command	Items, Partner Items Items, Partner Items Time-slot Command-Id	Accept *items* from a *sender*, and *items* already in stock, and transform them into *items* to be dispatched to a *receiver*, and *items* to keep in stock, in a *time-slot*. *Identification* for future reference.
Progress- Report	Progress Command-Id Items, Partner Items Items, Partner Items Time-Slot	Report about the *progress* of the execution (possible parameters are 'completed', 'busy', or 'waiting'.) of a previously-received *command*. Which *items* have been accepted from a *sender*, and together with *items* already in stock been transformed into *items* dispatched to *receivers*, and *items* kept in stock, in a *time-slot*.
Inventory-Indication	Items Time-Stamp	Indication of the *items* available and not yet committed at a certain *moment* in time.
Notes:	The parameter 'Items' also defines the number of items. The parameter 'Partner' can identify another Cell/Line, customer, or supplier. It also defines the location of the exchange of items.	

Table 4.1: Interactions Of Factory Controller And Cell/Line

4.5 Cell/Line–Cell/Line Interactions

Table 4.2 contains interaction primitives that can be used to model a Cell/Line Controller that negotiates with other Cell/Line Controllers or Customers and Suppliers to exchange items.

- The Item-Exchange-Agreement can be used to model the agreement on the parameters of the item exchange: the items to be exchanged, location of exchange, sender and receiver of items, time of exchange.

- The Item-Exchange-Progress can be used to confirm that items have actually been exchanged.

Interaction Primitive	Arguments	Explanation
Item-Exchange-Agreement	Items	Agreement to exchange *items*
	Location	can be exchanged at a *location*
	Partner	from a *sender* to
	Partner	a *receiver*
	Time-Stamp	at a point in *time*.
	Agreement-Id	*Identity* for further reference.
Item-Exchange-Progress	Progress	Report about the *progress* of item exchange
	Agreement-Id	per earlier *agreement*.
	Item	*Items* that have been exchanged.
Notes:	'Item' defines types and numbers of items.	

Table 4.2: Interactions Of Cell/Lines To Negotiate The Exchange Of Products

Table 4.3 contains interaction primitives that can be used to model the actual exchange of raw materials and products.

4.6 Decomposition of a Factory Controller

A Factory Controller's decision to execute certain jobs relates directly to the commands to the Cell/Line. Reversely, feed back from the Cell/Line about the execution and the job lead times directly updates the knowledge needed by the Factory Controller

Interaction Primitive	Arguments	Explanation
Product-Exchange	Products Sender Receiver	Exchange of *products* from a *sender* to a *receiver*.

Table 4.3: Interactions Of Cell/Lines To Exchange Products

to decide which jobs to execute. We cannot identify independent control tasks of a Factory Controller and therefore will not further decompose the Factory Controller. All interactions with Company Controller and Cell/Line directly relate to its essential task, the control of inventory points.

Note that we do not further decompose the Factory Controller to identify planning and control tasks, which are to be described in the Reference Model. However, if we were to decompose for other reasons, we might still be able to further decompose the Factory Controller. If we were interested in performance, for example, we could try to identify which subtasks of the Factory Controller could be executed in parallel.

We will discuss the decomposition of the Cell/Line in the subsequent chapters. We may identify Cell/Line components that we cannot decompose further, like the Factory Controller. In such cases, we will not explicitly justify why we do not decompose the components.

Chapter 5

Decomposition of a Cell/Line

We described the interactions of a Cell/Line in chapter 4. In this chapter, we analyse these to determine which tasks a Cell/Line should execute. We then decompose the Cell/Line into components that can execute these tasks.

5.1 Analysis of Cell/Line Interactions

A Factory Controller instructs a Cell/Line to execute jobs [1], i.e. to process items in certain time slots. The number and type of jobs may vary over time. A Cell/Line has a limited capacity and should therefore assess the feasibility of jobs and report to the Factory Controller whether it can execute them in the required time slots.

A Cell/Line needs resources to process items. It has to determine when these resources should be applied to manufacture specific items so as to ensure that jobs are executed in the allotted time slots. We will discuss these resources and their use. Let us first determine whether a Cell/Line should have one or multiple resources, whether they should realise of one or multiple job types, and whether they should execute a job entirely or partly. We tentatively define an 'operation' as the processing done by a resource, and will give a more accurate definition later.

Resources Executing Several Operations. In general, it would be preferable if a Cell/Line had resources that could execute several operations instead of just one operation. The point is that a Cell/Line may have to execute several jobs and should

---

[1]We use the word 'job' exclusively to refer to a transformation of items to be executed by a Cell/Line if requested by a Factory Controller.

be better able to utilise the resources for various combinations of jobs if, indeed, these resources can be used to realise several jobs.

Several Operations to Realise a Job. Resources preferably execute operations that realise a job partially rather than entirely. The reason is that such resources can be simpler from a technological point of view. Compare resources that only place components on a Printed Circuit Board, or resources that only solder them, or resources that only inspect them with a hypothetical resource that does it all, i.e placing, soldering, and inspecting.

Another advantage of a Cell/Line with resources that execute a job only partially is that the Cell/Line can execute a relatively large variety of jobs with a relatively small number of resources if these execute their operations in different sequences. The number of resources and operations is relatively small; the number of potential combinations thereof is relatively large.

Scheduling of Resources. Summarising, a Cell/Line needs multiple resources that can each do a portion of the processing required for several jobs. Given the resources and the jobs to be executed, a Cell/Line has to allocate operations to resources and to determine when those operations should be executed.

A 'schedule' is a plan that states which operation should be executed, when, and by which resource.

'Scheduling' is the generation and execution of a schedule.

Scheduling of operations is a very complicated task. Typically, one could evaluate very many different combinations, sequences, and allocations of operations to find a schedule that gives the Cell/Line a reasonable performance.

A 'part' is the result of an operation consisting of a composition of material and possibly some information about the material.

Note that both 'part' and 'item', which we defined in Section 4.1, refer to raw materials, partially processed materials, or products. We explicitly use 'item' to denote the input for and output from a job executed by a Cell/Line, and 'operation' to denote the input for or output from an operation executed to realise a job.

The schedule determines when parts are processed or waiting, and when resources are used or idle. Scheduling therefore affects the throughput of parts, time needed to complete jobs, and the utilisation of resources. A Cell/Line needs the ability to schedule operations to exploit the flexibility and efficiency potentials that exist because it has multiple resources that can execute several operations.

A Cell/Line could also use scheduling capabilities to assess the feasibility of a job proposed by its Factory Controller: it tries to develop a schedule to determine whether it can start and finish a job in the time slots outlined by the Factory Controller. An initial feasibility based on given and projected workload, can thus be shown and reported to the Factory Controller. However, the Cell/Line may be forced to re-schedule during the actual execution if new jobs are requested or if the resources do not perform as expected.

Operations as Pieces of Work to be Scheduled.   We identified operations as the activities of resources to be scheduled by a Cell/Line. All pieces of material processing that can be usefully scheduled should be identified as operations. In other words, an operation is a unit for scheduling that cannot be split into smaller steps that can be scheduled to improve the Cell/Line performance. If it would make sense to schedule these smaller steps, then they should be identified as operations.

> **Example.** Operation X takes parts A and B to produce part C. To pro-
> duce C, A and B have to be moved, heated, cooled, and moved again.
> These activities always take the same amount of time and should take
> place sequentially without intermission. Due to these constraints, it makes
> no sense to view these movement, heating, and cooling activities as sepa-
> rate operations. Scheduling them separately would not result in additional
> possibilities for improving the performance of the Cell/Line.

An 'operation' is a material processing activity that can be scheduled in a useful manner and that is considered an indivisible entity for scheduling because it cannot be split into material processing sub activities that can be scheduled in a more useful manner.

An 'operation lead time' is the time a given resource needs to execute an operation, once all parts needed for the operation have arrived at the resource.

We conclude that a Cell/Line should schedule operations. Let us now investigate which information a Cell/Line needs for scheduling.

Relevant Data for Scheduling.   The following data should be taken into consideration when developing a schedule:

• the jobs to be executed, and their time slots;

- the availability, capacity, and capability of resources;

- the availability of parts; and

- the costs and precedence relations of operations.

Some of these data tend to vary slowly during the time a Cell/Line has to execute a job. Consider, for example, the:

- precedence relations of operations, which depend on the design of parts;

- costs of operations, which primarily depend on the technological implementation of the resources and material characteristics of parts; and

- capacity and capability of resources, which primarily depend on the technological implementation of the resources and material characteristics of parts.

Because they change relatively slowly, these data can be viewed as given constraints on a schedule.

That does not hold for the jobs to be executed and their time slots:a Factory Controller may request the execution of a job at any time.

Similarly, the availability of parts and of the resources cannot be viewed as given constraints during the generation and execution of a schedule. They depend on the timing and allocation of operations, and therefore on the generation and execution of a schedule. Moreover, resources may fail to produce the required parts so that it is important that the actual availability of parts and resources is known when generating or executing a schedule.

Conclusion. A Cell/Line has to consider the jobs to be executed, schedule operations to realise these jobs, and to monitor whether execution of the schedule results in the desired availability of parts and resources. We concluded earlier that an operation should be viewed as a unit in a schedule; splitting the operations has no effect on the performance of a Cell/Line. It is therefore not important to know how an operation is executed, which sub tasks realise an operation, to be able to schedule it. Consequently, the task to generate and execute a schedule can be separated from the task to execute operations. Hence, we have identified the following, relatively independent, globally defined Cell/Line tasks:

- **Scheduling of Operations:** determine which operations on parts should be executed, by which resource, and when; and

---

Scheduling of Operations

Knowledge:

- how an item can be produced by executing operations on parts. This implies knowledge of precedence constraints of the operations, operation lead times, costs of operations, and capacity, capability, reliability of resources that execute operations; and

- of heuristics and algorithms to develop schedules.

Execution of Operations

Knowledge:

- how to change physical properties of parts so that they are transformed as required.

Figure 5.1: Knowledge Required For Cell/Line Tasks

- **Execution of Operations:** accept parts, execute operations on them, and dispatch the processed parts.

These tasks are essentially different. The scheduling task deals with combinatorics of satisfactorily scheduling operations to realise varying numbers and types of jobs. The execution task is technological in nature. Further, the scheduling task deals with names of parts without knowing the physical characteristics of the material to which these names refer, whereas the execution task requires knowledge of their physical characteristics. Figure 5.1 depicts the different kinds of knowledge required for the different tasks.

In the next Section, we describe how we can decompose a Cell/Line into distinct components that execute the tasks described above.

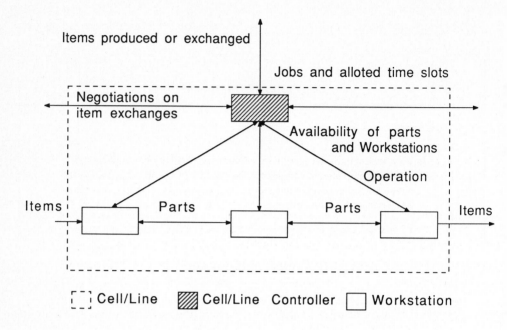

Figure 5.2: Decomposition Of Cell/Line Into Cell/Line Controller And Workstations

5.2 Cell/Line Controller and Workstations

In compliance with the development strategy discussed in Section 3.4, we assign the independent Cell/Line tasks, identified in Section 5.1, to distinct components, i.e. a 'Cell/Line Controller', which coordinates the execution of operations, and 'Workstations', which execute operations. Figure 5.2 illustrates our proposal for the organisation of a Cell/Line.

One Cell/Line Controller. A Cell/Line Controller receives commands to execute jobs, determines which operations should be executed to realise those jobs, schedules the operations, and commands Workstations to execute the operations so that the schedule is realised. The Workstations execute operations and exchange parts according to the Cell/Line Controller's commands and report to the Cell/Line Controller which operations they have executed.

In general, the information about the availability of all Workstations and all parts may be needed to develop a schedule. A Cell/Line therefore has only one Cell/Line Controller, coordinating all the Cell/Line's Workstations.

Multiple Workstations.   We concluded earlier that there may be several resources, or Workstations, which can execute operations.

Commitments on Time Slots.   The time it takes a Cell/Line to realise a job depends on the way it schedules operations. The schedule will be different each time the Cell/Line has to execute a different combination of jobs. Hence, the Factory Controller cannot easily and accurately predict the time slots in which a Cell/Line will execute the jobs. The Cell/Line Controller therefore assesses the feasibility of time slots proposed by the Factory Controller and notifies the Factory Controller of the feasibility.

As distinct from jobs, the time it takes to execute operations depends primarily on technological factors, and can be predicted with a reasonable accuracy. A Cell/Line Controller therefore does not need to seek a commitment from a Workstation telling that the latter will execute an operation in a certain time slot or before a due date.

Item Exchange.   A Factory Controller can request Cell/Lines to exchange items in certain time slots. The Cell/Line Controllers of those Cell/Lines negotiate with each other the location of the item exchange and the exact timing with in the time slots. They can instruct their Workstations to actually exchange them. The Cell/Line Controllers report to the Factory Controller that they have exchanged the items.

A similar mechanism can apply when a Factory Controller orders a Cell/Line to exchange raw materials or products with Suppliers or Customers. The Cell/Line and Customer or Suppliers express their preparedness to exchange raw materials or products, agree on the location, etc.

Workstations should be able to exchange the control of parts. The Workstations that exchange parts should manage the use of the space where the parts are exchanged. A Workstation that passes control of a certain space to another Workstation should, for example, refrain from reclaiming it before the other Workstation has reported that it has taken away the part.

Grouping of Workstations in Cell/Lines.   As described in Section 4.2, Cell/Lines should be able to execute their jobs independently from each other: the capacity of one Cell/Line to execute jobs should not be affected by the execution of jobs by another Cell/Line. In reality, however, Cell/Lines do affect each other's capacity since they exchange parts. Nevertheless, the required independence can be realised if Cell/Lines exchange parts only sparingly. The Workstations of a Cell/Line should form 'self-contained' groups, which exchange parts amongst themselves much more frequently than with Workstations of other groups. 'Much more frequently' is a qualitative judgement, of course. The idea is that a Cell/Line's execution of a job impact primarily its own Workstations, since these form a a self-contained group, and much less the Workstations of other Cell/Lines.

Observational Equivalence. We have identified a Cell/Line Controller and Workstation as the components of a Cell/Line and defined their global tasks. We can develop complete functional specifications for systems that execute the Cell/Line Controller and Workstation tasks. As discussed in Section 3.5, we require that a parallel composition of specifications of a Cell/Line Controller and Workstations be observational equivalent to a specification of a Cell/Line.

More formally, let,

process *Cell/Line*[cu,su,int,int1] : **noexit** :=
      (* specification of complete functional behaviour of a Cell/Line. *)
endproc

be a LOTOS-like specification of a Cell/Line as a process that interacts at gate:

- 'su' with Suppliers to exchange raw materials; and

- 'cu' with Customers to exchange products; and

- 'int' with a Factory Controller to receive commands to execute jobs; and

- 'int1' with other Cell/Lines to exchange items.

Figure 5.3 depicts a Cell/Line as a parallel composition of processes '*Cell/LineCon-troller*' and '*Workstations*'.

Let,

process *Cell/LineController*[cu,su,int,int1,int2] : **noexit** :=
      (* specification of complete functional behaviour
      of a Cell/Line Controller. *)
endproc

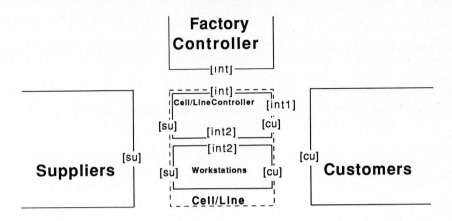

Figure 5.3: A Process Representation Of A Cell/Line As A Parallel Composition Of A Cell/Line Controller and Workstations

be a specification of a Cell/Line Controller as a process interacting at gate:

- 'cu' with Customers to determine where the products should be exchanged;

- 'su' with Suppliers to determine where the raw materials should be exchanged;

- 'int' with the Factory Controller to establish which jobs it should execute;

- 'int1' with other Cell/Line Controllers to determine where items should be exchanged; and

- 'int2' with Workstations to request them to execute operations.

Let

process *Workstations*[cu,su,int2] : **noexit** :=
        (* specification of complete functional behaviour of Workstations. *)
endproc

be a specification of Workstations as a process interacting at gate 'int2' with a Cell/Line Controller to receive commands to execute operations, at gate 'cu' with Customers to dispatch products, and at gate 'su' with Suppliers to accept raw materials.

We then require that:

Cell/Line[cu,su,int,int1] ≈ **hide** int2 **in**
(
Cell/LineController[cu,su,int,int1,int2] |[int2]| *Workstations*[cu,su,int2]
).

Similarly, we require observational equivalence of the specification *Cell/Lines*, discussed in section 3.5, and a parallel composition of an arbitrary number of specifications *Cell/Line* discussed above. In a LOTOS-like manner, this is expressed as:

Cell/Lines[cu,su,int] ≈ **hide** int1 **in**
(

Cell/Line[cu,su,int,int1]
|[int1]|
Cell/Line[cu,su,int,int1]
|[int1]|

..

..

|[int1] |
Cell/Line[cu,su,int,int1]

(*Here, we do not consider possible parameters for *Cell/Line*. *)

)

Figure 5.4 illustrates *Cell/Lines* as a parallel composition of an arbitrary number of processes *Cell/Line*.

If all the above observational equivalence relations hold as well as those in section 3.5, then the parallel composition of a specification of:

- a process *FactoryController*, and

- one or more processes *Cell/LineController*, each in parallel composition with

- a process *Workstations*.

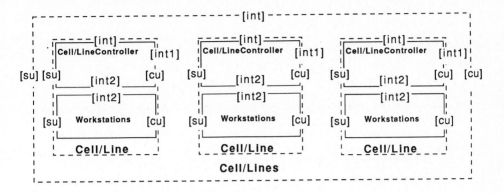

Figure 5.4: **A Process Representation Of Cell/Lines As A Parallel Composition Of Multiple Instances Of A Cell/Line**

is observational equivalent to the specification of a process *MPCS*.

Henceforth, we will implicitly require that complete functional specifications of MPCS components adhere to observational equivalence relations as discussed above; we will no longer explicitly define these relations.

In the following Section, we describe some practical situations with Cell/Line Controllers and Workstations.

5.3 Modeling Real-Life Situations

In this Section, we describe how Cell/Line Controllers and Workstations cooperate to realise Cell/Lines. We describe some general tasks that a Cell/Line Controller may be required to execute and some typical Cell/Line Controller—Workstation interactions. Finally, we make some remarks about the scheduling procedures of Cell/Line Controllers, and compare these with procedures available in literature.

General Requirements for Cell/Line Controllers. A Cell/Line Controller may have to accomplish a variety of tasks. We developed specifications for a Cell/Line Controller [21], for example, that can:

- Ensure that the flow of parts such as screws, resistors, stickers, picture tubes, and coils, is coordinated with the flow of parts with which they are to be assembled;

- Determine the sizes of batches and series of parts processed by Workstations;

- Determine which of the parts waiting for a Workstation should be processed first;

- Determine when the execution of a job should start;

- Determine whether parts that fail to successfully pass tests executed by Workstations should be repaired by repair Workstations;

- Ensure that 'buffer Workstations' contain sufficient parts; and

- Schedule Workstations.

Let us now review some typical scenarios.

Scheduling.   We have discussed that a Cell/Line Controller has the task to schedule the operations for Workstations. Scheduling tends to be a computationally intractable problem [46]. A Cell/Line Controller therefore often executes heuristic rules to develop a schedule. These rules can be executed quickly and result in a schedule that is reasonable, though not optimal, in many situations. One such a heuristic is the Shortest Processing Time (SPT) rule. This rule tells to select the next part for processing that has the shortest processing time at a Workstation [105].

Workstations are often arranged in 'job shops'. Workstations with similar capabilities –typically expensive Workstations such as for component insertion or metal cutting– are grouped together. An important scheduling objective is to maximise the utilisation of the Workstations; the waiting times of parts is of secondary concern. Parts have to wait close to the group of Workstations and are moved to the first available Workstation. Job shops are therefore commonly associated with high Work-In-Progress levels.

Workstations are also often arranged in 'flow shops' [19,91,94]. Workstations with distinct capabilities are arranged in a line, and subsequently execute operations required to realise a job. One Workstation processes a part, passes the resulting part to the next Workstation, which processes it and passes it on. There are variations on this concept, where parts may skip some Workstations, or have to visit one of some identical Workstations. References [11,69,109] describe scheduling heuristics for flow shops.

The scheduling of flow shops is typically more constrained than for job shops. The major practical scheduling issues are the sequence of parts entering the line and the batch sizes. Set ups of Workstations, which take time, and operations with different operation lead times, which tend to cause blocking and starvation of Workstations, are the major costs incurred by a schedule.

References [82,88,53] review scheduling objectives and constraints used in practice and discuss how these differ from those described in literature. In particular Reference [88], mentions the following requirements of industrial scheduling procedures, which are hardly addressed by the scheduling literature:

- Industrial scheduling is really re-scheduling, i.e. modifying existing schedules to accommodate for new jobs, new job priorities, or unscheduled disruption in the production. Scheduling procedures must therefore be capable of accommodating many initial states, i.e. the relevant states of a Cell/Line when a rescheduling occurs;

- Scheduling procedures must explicitly determine batch sizes given scheduling objectives, jobs, and, possibly sequence dependent, set-up costs;

- Scheduling procedures must be able to exploit possible alternatives in the combination and sequence of operations; and

- Scheduling procedures often have to satisfy multiple, conflicting, and sometimes non-commensurate constraints and performance objectives.

Materials Requirements Planning.   In practice, MRP systems are often used as scheduling tools. They assume a rigid relation between items to be manufactured, operations to be executed for their manufacture, and the moments in time that the operations have to be executed. They ignore the fact that operation lead times may vary and that the time between operations depends on the actual availability of Workstations and therefore on the Work-In-Progress.

These relations are expressed by a 'Bill Of Material'. The MRP systems use this BOM to translate the demand for items into a time-phased demand for parts, compare this demand with the inventory, and planned replenishment and depletion of this inventory, and then determine which parts have to be transformed and when.

Typically different departments make these parts. Each has a scheduler who can schedule the local Workstations to produce them. However, if the MRP systems schedule parts to be produced by departments and not for the local Workstations, they allow these departments to develop sub optimising departmental schedules.

Moreover, the MRP systems are very rigid. They do not optimise the sequence of operations. They make worst-case assumptions about the operation lead times, whereas these lead times should be reduced by proper scheduling. We may conclude that MRP systems, used as scheduling tools, are simple but fail to actually schedule operations to achieve objectives so as to minimise job lead times, optimise throughput, or optimise Workstation utilisation.

Just-In-Time Production. "Just-In-Time", or "JIT" refers to situations where Workstations dispatch a part only when it is fairly sure that a downstream Workstation actually needs it at the moment that it will arrive [19,79,91,94]. Hence, it is avoided that parts are queued in front of a downstream Workstation.

The Workstations contain at least one part, which they can quickly dispatch to respond to the demand of a downstream Workstation. JIT systems therefore contain some inventory, which facilitates a quick response as a kind of catalyst. In practice, there is more inventory than the minimum of one part per Workstation to prevent that a Workstation runs out of stock if demand increases in frequency or volume [2].

JIT systems can be modeled by a Cell/Line Controller and Workstations. The Cell-/Line Controller receives status information from Workstations and therefore knows when a Workstation is ready to start the execution of an operation, and when it needs parts. It can then command an upstream Workstation to dispatch the needed parts.

Modeling the control of inventory levels is less trivial. Some inventory is not strictly the result of a Cell/Line's scheduling of operations to execute a job. Rather the inventory is created to facilitate the execution of these operations. There will be some amount of inventory in a JIT system even when the Cell/Line Controller has no jobs to execute. When it has to execute a job, the Cell/Line Controller commands a Workstation to dispatch a part of the available inventory and commands other Workstations to replenish the dispatched part. We will model the creation of this type of inventory in Section 13.2.

Buffer Stock. A Cell/Line Controller also has the responsibility to determine which parts a Workstation should contain as buffer stock. Buffer stocks serve to compensate for differences or variations in the output of Workstations, for example, caused

---

[2]Note that JIT systems as described above are only feasible in situations with relatively stable, repetitive demand for a low variety of parts; otherwise the investments in inventory would be too high.

by failures or unbalanced loading of Workstations.

Parts in a buffer can be passed to downstream Workstations if an upstream Workstation temporarily fails to produce a sufficient amount of these parts. The buffer stock then prevents a starvation of the downstream Workstation. Similarly, parts from an upstream Workstation can be put in a buffer stock if a downstream Workstation temporarily fails to accept them. The buffer stock then prevents a blocking of the upstream Workstation.

Both, these buffers and the inventory in JIT systems discussed above, are not created strictly as a result of the execution of jobs, but rather as a means to facilitate the execution of jobs so as to realise a certain performance. We will model the creation of this type of inventory in Section 13.2.

Transport System. We can model a transport system as a particular, specialised Workstation. Its basic operation is to accept parts from a Workstation at a certain location and pass them to a Workstation at another location. The Cell/Line Controller commands a Transport System from where to where a part is to be transported and the Transport System determines how it executes its operation, i.e. how it transports the parts and along which route [12].

Figure 5.5 shows a situation in which a Cell/Line Controller sends a part to a point en route to multiple Workstations and, subsequently, from that point to one of the Workstations. In this way, it can postpone its selection of a destination Workstation, and select one that is available when the part arrives at the intermediate point. Note that the Transport System cannot select a Workstation since it has no knowledge of the Workstation capabilities.

We can model such a situation by two separate Transport Systems that exchange parts. The Cell/Line Controller commands the first Transport System to accept a part and to exchange it with the second at the bifurcation. When it is ready to select a Workstation, the Cell/Line Controller tells the second Transport System to accept the part at the decision point and to transport it to the selected Workstation.

Figure 5.6 shows a transport system with a loop. Typically, the number of parts in the loop should stay within certain limits to avoid congestion and, consequently, a degradation of throughput.

Congestion is not affected by the technology of transportation but by the number and sequence of transport operations. This indicates that the Cell/Line Controller rather than a Transport System should control the number of parts in the loop. Figure 5.7 illustrates a possible scenario. We model three Transport Systems, i.e. one to move parts:

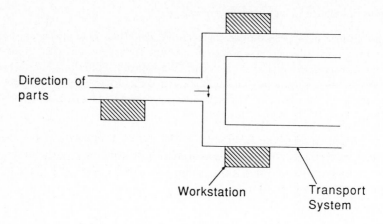

Figure 5.5: Selection Of Next Workstation When Parts Are At The Bifurcation

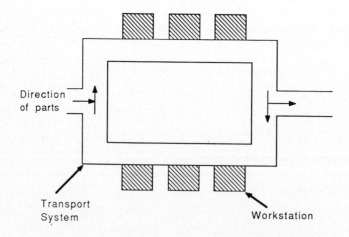

Figure 5.6: Transport Loop

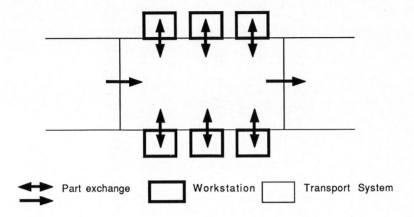

Figure 5.7: Loop Control

1. to the loop;

2. into and out of the loop, and to and from the adjacent Workstations; and

3. away from the loop.

The Cell/Line Controller commands the Transport System that controls the loop, to accept parts from the Transport System that brings the parts, or to accept parts from the adjacent Workstations. Thus, the Cell/Line Controller determines which parts are admitted to the loop. The Cell/Line Controller can further command the Transport System that controls the loop, to pass parts to the adjacent Workstations, or to the Transport System that takes the parts away. Thus, the Cell/Line Controller determines which parts leave the loop. Concluding, the Cell/Line Controller can determine the number of parts in the loop and keep this number within certain limits.

Parts and Tools.   We defined a part as material and associated information resulting from an operation. An operation may lead to the assembly of parts, or disassembly, or both. In other operations, such as test operations, the material of parts is not changed but information about the material is generated.

When a Cell/Line Controller commands a Workstation to execute an operation, it should define the parts that are involved, i.e. the parts that the Workstation:

- must accept from other Workstations and use for the operation;

100

- has already available in its own domain and must use for the operation;

- has to produce and dispatch to other Workstations; and

- has to produce and keep within its own domain;

From a control perspective, a tool can be treated as a part. It is also a resource needed to execute an operation. The Cell/Line Controller can command a Workstation to accept or dispatch a tool, or to use it for a certain operation. Tools that are not distructed during an operation can be viewed as parts that are to be used and to be produced as well.

Part Data.   Workstations may also have to exchange information related to parts.

'Part data' is data related to parts and exchanged by Workstations.

Although part data is related to a part, it does not need to visit the same Workstations as the part itself.

> **Example.** A Workstation may have results of a test on a part that is relevant for the Workstation that has to repair the part. On command from the Cell/Line Controller, it passes these data as part data directly to the repair Workstation. It passes the part itself, on command from the Cell/Line Controller, to a Transport System, which will be commanded to bring it to the repair Workstation.

From the perspective of the Cell/Line Controller, parts and part data are treated in the same way. The Cell/Line Controller commands a Workstation to produce the physical parts and the part data, and to pass them to other Workstations.

Layout of Workstations.   We have argued that the relative frequency with which Workstations exchange parts should determine whether they belong to the same Cell/Line. In practice, however, it is often assumed that the geographical configuration of Workstations, the layout, determines whether they are grouped together in Cell-/Lines [86].

We have seen situations as illustrated in Figure 5.8, where Workstations were allocated to the same Cell/Line because they were located in the same room. However, the Workstations within one room did not exchange parts; parts visited Workstations in different rooms. Hence, it would make more sense to allocate the Workstations that exchange parts to one Cell/Line, despite the fact that they are in different rooms.

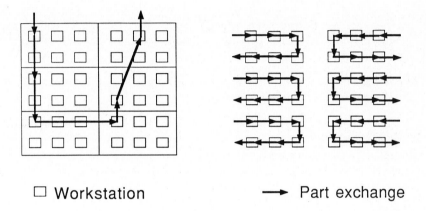

☐ Workstation   ⟶ Part exchange

The left picture shows Workstations that are grouped together for environmental reasons. They have, for example, different requirements regarding the cleanliness of the air, like in Integrated Circuit manufacturing facilities.

The right picture shows Workstations that are grouped together because they exchange parts frequently.

Figure 5.8: Self-Contained Groups Of Workstations Exchanging Parts, Whether They Are Geographically Close Or Dispersed.

5.4 Cell/Line Controller–Workstation Interactions

Table 5.1 contains interaction primitives [3] that can be used to model Cell/Line Controller and Workstations that:

- Agree on the execution of an operation. The operation is defined by an Operation-Command which tells which parts are to be accepted, processed, dispatched, etc. Repeat operations can be ordered by multiplying the number of parts.

- Reports about was has actually happened as opposed to what was commanded. The 'Progress-Report' primitive can model this.

  There may be several Progress-Reports referring to the same Operation-Command with its Command-Id.

  Note that the parts mentioned in a Progress-Report may differ from those mentioned in the ordered Operation-Command. Defects may have been produced, for example. Or, more parts have been used than ordered.

  A 'completed' Progress-Indication will be issued when an operation is finished or has been canceled by the Factory Controller.

  A progress parameter 'busy' can be used to report about the parts that have been exchanged already.

  A progress parameter 'waiting' indicates that Operation-Command cannot be executed because parts are not there. The part parameters are used to indicate which parts are available, have already been accepted, etc.

5.5 Cell/Line Controller–Cell/Line Controller Interactions

The Cell/Line Controllers negotiate the exchange of parts between the Cell/Lines and command their Workstations to actually exchange the parts.

The interaction primitives in Table 4.2 were given to model the interactions of Cell-/Lines to negotiate the exchange of parts. These can be used as well to model the negotiations by Cell/Line *Controllers* since these conduct the negotiations.

---

[3]Appendix A describes the meaning and role of interaction primitives.

Interaction Primitive	Arguments	Explanation
Operation-Command	Parts Stock-Parts Parts Stock-Parts Command-Id	Command to accept *parts* and together with *parts* in stock transform them into *parts* to be dispatched and *parts* to keep in stock, Command *identity* for future reference.
Progress- Report	Progress Command-Id Parts Stock-Parts Parts Stock-Parts	Report about the *progress* (possible parameters are 'completed', 'waiting', or 'busy'.) of a *command*: which *parts* were accepted, and which *parts* in stock were transformed into *parts* dispatched and *parts* kept in stock.
Notes:	'Parts' defines the type and number of parts to be exchanged as well as their sender or receiver and the location of part exchange. 'Stock-Parts' defines the type and number of parts.	

Table 5.1: Interactions Of Cell/Line Controller And Workstation

5.6   Workstation–Workstation Interactions

Table 5.2 contains interaction primitives that can be used to model Workstations that can transfer the control of parts.

- The Part-Control-Exchange primitive models an agreement of two Workstations to exchange the control of a part at a certain space. The implies that the Workstation relinquishing control should withdraw from that space and transfer the control of that space to the Workstations that will accept the part.

- The Space-Control-Exchange primitive models an agreement of two Workstations to exchange the control of a space without transferring a part.

Interaction Primitive	Arguments	Explanation
Part-Control-Exchange	Parts Partner Partner Space	Agreement to exchange *parts* from a *sender* to a *receiver* in a *space*.
Space-Control-Exchange	Space Partner Partner	Agreement to the control of a certain *space* from the *Workstation* currently controlling it to a *Workstation* that is going to control it.

Table 5.2: Interactions Of Workstations

Chapter 6

Decomposition of a Workstation

We described the interactions of Workstations in Section 5.4. Here, we analyse these to determine which tasks a Workstation should execute.

6.1 Analysis of Workstation Interactions

A Cell/Line Controller commands Workstations to exchange parts and to execute operations on them so that they are transformed, inspected, stored, or transported. A Workstation may be able to execute several operations and to execute multiple operations concurrently.

A Workstation must execute such 'processing steps' as grinding, polishing, colouring, heating, or displacing parts to change their physical characteristics as is required to complete an operation. In Section 5.1, we defined an operation as a unit for scheduling, which is not further decomposed into material processing activities that should be scheduled. A Workstation, therefore, does not schedule processing steps to improve its performance. Instead, it executes them in an order that depends on the arrival of parts and the availability of resources to execute the processing steps.

A Workstation needs resources to execute the required processing steps. However, operations often require processing steps that are entirely different from a technological perspective, whereas a typical resource can only economically execute processing steps that are technologically related. An operation may imply, for example, that parts be moved, heated, and split, which all require technologically different processing steps executed by different resources. Workstations may therefore have to coordinate multiple resources. Each of them can execute technologically related processing steps in an economically feasible manner.

An 'object' is the result of the execution of a processing step, consisting of material.

Note that 'object' denotes an intermediate stage of a 'part', which according to the definition in section 5.1, refers to the material exchanged by Workstations as input for or output from an operation.

Objects have certain physical characteristics, like location, shape, temperature, and roughness, which are changed by processing steps.

To execute an operation, a Workstation should take into account the:

- availability and location of objects;

- processing steps required to execute an operation and their precedence relations; and

- availability and capabilities of resources that execute processing steps.

Some of these constraints are quasi static during the time needed to execute an operation. The capability of resources, for example, depends on their physical construction, and changes slowly. Similarly, the processing steps required to execute an operation depend largely on the capabilities of available resources and the design of parts, which change slowly.

Consequently, a Workstation can treat the capabilities of resources and the required processing steps as given constraints when it has to execute an operation. It should keep track of the objects and ensure that the right resource executes the right processing steps when the objects and resources are available. It is not important to know how process steps are executed to coordinate their execution.

We have therefore identified relatively independent Workstation tasks, which are described below.

- **Coordinate Processing Steps:** determine which resource should execute which processing step on what object so that operations are realised; and

- **Execute Processing Steps:** process objects to change their physical characteristics and report about these characteristics.

These tasks are essentially different in purpose. The coordination tasks serves to bridge the gap between the operations, the logistic units to be scheduled, and the available technological capabilities of the resources. The execution task serves to realise these capabilities. Figure 6.1 illustrates that both tasks require different types of knowledge.

Coordinate Processing Steps

Knowledge:

- which processing steps are required to execute operations, in what order, and on which objects;

- to recognise and keep track of objects;

- to coordinate the exchange of objects.

Execute Processing Steps

Knowledge how to:

- change physical characteristics of objects;

- recognise physical characteristics objects.

Figure 6.1: Knowledge Required For Workstation Tasks

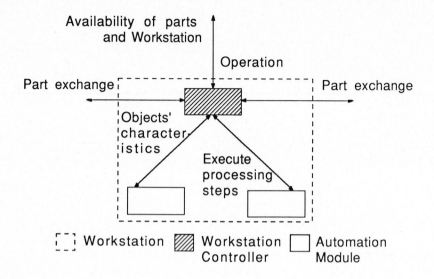

Figure 6.2: A Workstation Decomposed Into A Workstation Controller And Automation Modules

6.2 Workstation Controller and Automation Modules

In compliance with the development strategy discussed in Section 3.1, we assign the relatively independent Workstation tasks identified in Section 6.1 to distinct components. A 'Workstation Controller' [20] has to coordinate processing steps, and an 'Automation Module' has to execute processing steps. Figure 6.2 illustrates our proposal for the organisation of a Workstation.

Single Workstation Controller. A Workstation Controller should see to it that objects undergo all the processing steps required to produce the parts ordered by the Cell/Line Controller. The overview of all objects being processed can best be obtained by a single Workstation Controller. We therefore propose that a Workstation have only one Workstation Controller.

Multiple Automation Modules. We already mentioned that an operation may require the execution of several, technologically distinct processing steps. The resources available for their execution can typically execute a only limited number

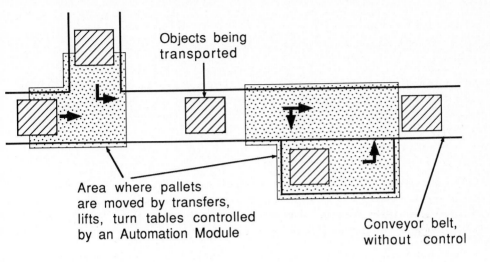

Figure 6.3: Multiple Automation Modules In A Transport System

The Workstation Controller of the Transport System determines the path of each object to the destination outlined by the Cell/Line Controller. Automation Modules execute movement steps at different nodes in the Transport System as commanded by the Workstation Controller. The movements at different nodes can be executed independently of each other so that each node can be controlled by a separate Automation Module.

of technologically related processing steps. A Workstation Controller may therefore need multiple Automation Modules, which can execute one or more technologically related processing steps. A Workstation Controller may also have multiple Automation Modules that execute technologically similar processing steps, provided that they can be executed independently of each other. Figure 6.3 illustrates an example.

There is no need for Automation Modules to exchange information with each other since their domains and processing activities can be coordinated entirely by the Workstation Controller.

Upon receipt of a command from the Cell/Line Controller to execute a certain operation, the Workstation Controller will wait for indications from other Workstations that the parts implied by the operation are available. Subsequently, it commands one or more Automation Modules to execute the required processing step, waits till they are ready, commands other Automation Modules, etc. Finally, it dispatches the parts to other Workstations. It reports to the Cell/Line Controller which parts it has dispatched, and which ones it has kept.

Automation Modules execute processing steps and return status to indicate that the steps have been executed, the failures that occurred, etc. Some Automation Modules will report to the Workstation Controller specific physical characteristics of objects, such as a bar code reading, or shape, or weight, etc. This enables the Workstation Controller to recognise objects, and to assess their quality.

Observational Equivalence.   As discussed in section 3.5, we can develop complete functional specifications of MPCS components on the basis of the definition of their global tasks. We could therefore develop such specifications for Workstation Controllers and Automation Modules.

We require that a parallel composition of specifications of a Workstation Controller and one or more Automation Modules be observational equivalent to a specification of a Workstation. This could be expressed formally in a way similar to the observational equivalence relations in section 3.5.

In the next Section, we describe some practical situations and the role of Workstation Controllers and Automation Modules in these situations.

6.3   Modeling Real-Life Situations

General Workstation Tasks.   A Workstation may have to accomplish a variety of tasks. We specified [18,20] and implemented [107], for example, a Workstation that can:

- accept parts from and dispatch parts to multiple other Workstations;

- recognise parts on the basis of their identification mechanisms such as bar codes, on the basis of their geometry or other physical characteristics, or on the basis of the information given by the workstation with which it exchanges them;

- execute multiple operations simultaneously;

- change tools or perform other set-up activities for an operation;

- maintain a local buffer of parts such as screws or electronic components;

- inspect objects during the operation, it may have to move objects, have to store objects, have to assemble or disassemble objects;

- collect part data (see Section 5.3);

- coordinate a variety of Automation Modules; and

- can honour requests to suspend or cancel the execution of operations.

A Workstation that has been commanded to execute multiple operations, will wait for parts to arrive, and execute those operations for which the parts are available. This means that the order in which it executes operations depends on the arrival of the parts.

However, a Cell/Line Controller may want to have more control over the timing of the operations in light of its responsibility to schedule the operations. It can, for example, first command a Workstation to store certain parts. After the Workstation has reported which parts it has stored, the Cell/Line Controller can select the highest priority operation for which the parts are available and command the Workstation to execute this one.

Further, the Cell/Line Controller may be able to command Workstations to suspend and continue the execution of operations [20] and thus influence the order of their execution.

Coordination of Automation Modules. A Workstation Controller may have to coordinate Automation Modules, with overlapping geographical domains. The Workstation Controller's coordination should prevent undue interference, for example, by keeping Automation Modules within certain geographical domains. Similarly, the Workstation Controller has to oversee that Automation Modules do not unduly interfere with objects. The domains of Automation Modules may also be mutually-exclusive in a non-geographical sense. A Workstation Controller may, for example, have to coordinate the generation of heat by one Automation Module and its distance from another Automation Module to prevent the other from being distructed by the heat.

Resource Management. The Workstation Controller is also responsible for 'resource management', the resolution of conflicting claims on Automation Modules, space, objects, and tools for the execution of multiple operations. Figure 6.4 illustrates an example.

112

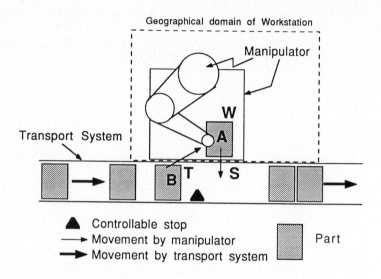

During the execution of an operation, the Workstation Controller commands the 'Manipulator' Automation Module to put object A at location W. The Manipulator is not needed for the next processing step to be performed on A and therefore becomes 'idle' after having put A on W.

Now Part B arrives. Suppose that the Workstation has been commanded to execute an operation on B. As a first processing step for the execution of this operation, the Workstation Controller should command the Manipulator to pick up B and to put it at location W [18]. However, it should refrain from giving the Manipulator this command despite the fact that it is idle because there is no room for B within the domain of the Workstation.

Trying to pick up B by the Manipulator would result in a dead lock: B cannot be put anywhere, and there is no free manipulator that can make room for B.

Figure 6.4: Resource Management

Parts Exchange. Workstation Controllers should interact to determine whether and when they exchange parts. They should regulate their access to the space that contains the parts to be exchanged. A Workstation Controller with the right to access a space can command one of its Automation Modules to put a part in or withdraw a part from that space.

> **Example.** Figure 6.4 illustrates two Workstations that exchange parts, i.e. a Workstation that contains a Manipulator and a Transport System. If the Transport System has brought a part destined for the Workstation to location T, it should interact with the Workstation to pass control of the part. Once the Workstation has accepted the control of the part, and removed it from T, it should return the right to use location T to the Transport System. A similar scenario applies if the Workstation has to return the part to the Transport System and put it onto location S.

Practical implementations of Workstation Controllers typically do not engage in the interactions to exchange parts as described above. Often, there is no need to manage space in order to exchange parts since the Automation cannot really interfere. One Workstation pushes a part onto a certain spot where the other takes it. The Workstation will temporarily stop pushing the part if it meets resistance because the spot is still occupied. Nevertheless, we feel that even in these simple cases a full implementation of the interactions to exchange parts would be useful. They greatly reduce the need for explicit identification of parts. The Workstation Controller dispatching a part can interact with the Workstation receiving the part and transmit the identification of the part. The receiving Workstation Controller does no longer need Automation Modules (like bar code readers) to explicitly identify the part. References [20,28,92] describe in some detail how Workstations interact to exchange parts.

Transport Systems. We can view a Transport System as a specialised instance of a Workstation. The Workstation Controller of a Transport System determines which movement steps should bring a part to the destination outlined by the Cell/Line Controller. It selects the route and therefore is often referred to as a 'routing controller' [8,12,28]. The Automation Modules execute the movement steps selected by the Workstation Controller. In case of pallet transport, the Automation Modules are implemented by lifts, turn tables, transfers, etc. In case of Automated Guided Vehicle transport, the Automation Modules are implemented by mobile carriers.

Practical Commands to Automation Modules. It is important to realise that Automation Modules can recognise and process *objects*. A system that can be commanded to move an object is an Automation Module; a system that can be commanded to move only itself is not.

A typical command to a robotic Automation Module would be to move an object with dimensions X and Y from location A to B. A particular Automation Module may not be sophisticated enough to process such concise commands. Instead, commands should be given like: open gripper, move to location A, close gripper to grasp object with dimensions X and Y, move to location B. Together, these commands convey the same information as the concise one.

In practice, commands to Automation Modules are often implemented by down loading programs. Instead of giving an Automation Module a command to place a component onto a Printed Circuit Board, a program is down loaded that, when executed, would have the same effect.

Identification of Workstations. It should be noted that one often assumes that a Workstation is implemented by a machine or human operator. Rather than focusing on the physical systems, we suggest to identify the operations first: which pieces of work should be scheduled so that products can be manufactured efficiently. The physical systems that are used to execute these operations are the physical implementation of a Workstation. These physical systems may comprise a single or multiple machines, may be small or big, with or without human operators, etc.

Implementation of Workstations. To illustrate the various comments about operations, part exchanges, and identification of Workstations, we consider the control of a rather complex system, illustrated by Figure 6.5.

The system can process parts by exposing them to laser beams. It can, for example, cut garments out of cloth. It consists of two robots, two tables, a mirror, and a laser. Each robot is used to put parts onto a table and to take them off this table. The mirror can be set in two positions to direct a laser beam to either one of the tables. Each table can be moved independently, and can be moved in the X and Y directions independently. A laser can be used to generate a laser beam with variable intensity. Depending on the position of the mirror, the beam will reach either table. The effect of the laser on a specific part depends on the part's material characteristics, the part's position on the table, the laser's intensity, and the table's position and velocity.

The laser has a physical control system to control its intensity. There is only one such a control system. In addition, each table and robot has its own control system. How should the physical control system of the laser be coordinated with those of the

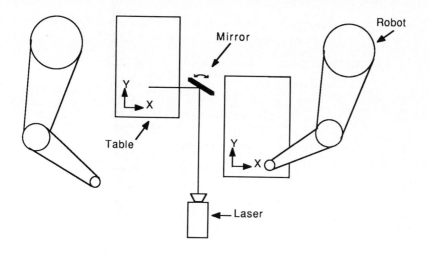

Figure 6.5: A Laser Processing System

tables and robots?

Let us assume that there is a Cell/Line Controller that determines which operations the system has to execute. Let us further assume that the laser processing of a part will not be interrupted to process another part.

We should first identify the operations involved and the Workstations that should execute them. Recall that operations are pieces of work to be scheduled and that Workstations should be able to execute operations independently of each other.

There may be good reasons to schedule the processing of parts for the two tables separately. Suppose, for example, that one knows for each specific part the time required to put it onto or off a table as well as the time required for processing it with the laser. One can now schedule the processing of parts to, say, minimise the idle time of the laser. To that end, one could try to sequence parts to the tables so that the laser is always processing a part on one table during the time a robot needs to put a part onto or off the other table. The following operations should therefore be scheduled: the laser processing of parts, the placement of parts onto the table, and the placement of parts off the table. Note that there is no need to schedule table movements or laser emission since we assumed that the processing of a part would not be interrupted once it was started.

The above operations can be executed by both robot-table combinations independently, although only the combination with access to the laser can expose the part to

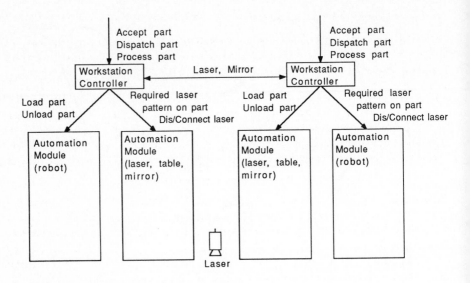

Figure 6.6: Abstract Model For The Control Of The Laser Processing System

laser beams. We can therefore view the system as consisting of two Workstations. Each Workstation contains a robot and a table, and they exchange the laser and mirror on command of the Cell/Line Controller. Thus, we view laser and mirror as tools.

Let us temporarily assume that each Workstation has an Automation Module to control the laser, irrespective of whether it has access to the laser. We can now model the system as illustrated by Figure 6.6.

Each Workstation Controller coordinates two Automation Modules. One Automation Module, which controls the robot, can be requested to put parts onto or off the table. The other Automation Module, which controls the laser, table, and mirror, can be requested to expose a part to a certain laser pattern part and will coordinate table movements and laser intensity. Note that table movement, laser intensity, and mirror position are technologically coupled since they all affect the impact of the laser on the part. Hence, they should not be viewed as independent processing steps and should be coordinated by one Automation Module.

Figure 6.6 illustrates that the Workstations can exchange the laser (and mirror). Notice, however, that they exchange the *control* of the laser, which does not imply that they necessarily displace the laser. In fact, they do not displace the laser but rather connect or disconnect it. They can command an Automation Module to do so. When the Automation Module is asked to disconnect the laser, it refrains from controlling it.

The model in Figure 6.6 does not seem to accurately reflect the fact that in reality there is only one control component to control the laser and mirror since the Figure shows two Automation Modules that can control the laser.

However, the model is correct as a Reference Model. A Reference Model defines the tasks executed by Workstation Controllers and Automation Modules, irrespective of whether some or portions of these Automation Modules are physically implemented by the same system. Note also that Figure 6.6 does not imply that two Automation Modules can concurrently control a laser since the laser itself is exchanged by the two Workstations as a critical resource.

Let us now develop a model that more closely reflects the physical system. It should have one MPCS component that controls the laser for laser processing on both tables. (We will focus on the laser and ignore the mirror). Let us first continue our reasoning and consider both Automation Modules that control a laser, irrespective of whether they are implemented on the same physical system. We decompose these Automation Modules to reveal more detail about their internal organization. As Figure 6.7 illustrates, and as the theory in the following chapters will show, we can decompose these Automation Modules into:

- an Automation Module Controller determining the required X and Y movements of a table, and the required laser intensity;

- three Device Controllers, which can execute commands from the Automation Module Controller to:

    1. realise a certain profile of the table's X position; and which in turn controls a motor; and

    2. realise a certain profile of the table's Y position, and which in turn controls a motor; and

    3. generate a laser beam with a certain intensity profile, and which in turn controls a physical laser, respectively.

The Workstation Controller can command the Automation Module Controller to expose a part to a certain laser pattern. The Automation Module Controller will coordinate the X and Y movements of the table and the laser intensity by issuing appropriate commands to its Device Controllers.

When, it is requested to disconnect the laser, the Automation Module Controller will refrain from sending commands to the Device Controller that controls the laser.

We will now consider the physical implementation and determine how, as in the real system, there can be a single control component that controls the laser. Figure 6.8

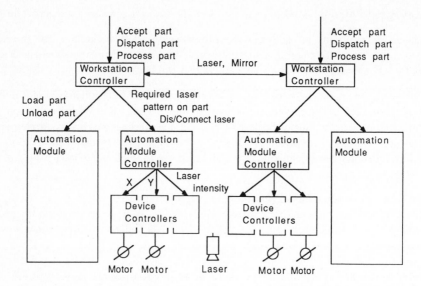

Figure 6.7: Decomposition Of Automation Modules That Control A Laser

illustrates an implementation model. The two abstract Device Controllers, each controlling their laser, are implemented by one physical Device Controller controlling one laser. The Device Controller is shared by two Automation Module Controllers. These will control it alternately: only the Automation Module Controller that belongs to the Workstation that got control over the laser will command the Device Controller of the laser.

The above example shows how we can conveniently model a complex system if we consider it at different levels of abstraction. It also illustrates how savings on the physical construction of a system, as the elimination of a laser, can be offset by a more complex control system.

6.4 Workstation Controller–Automation Module Interactions

Table 6.1 contains interaction primitives [1] that can be used to model an Automation Module that:

---

[1] Appendix A explains the purpose and meaning of interaction primitives.

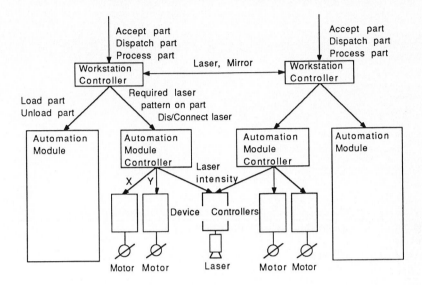

Figure 6.8: Model Of An Implementation Of The Laser Processing System

- Can be commanded to transform objects. The objects are defined by such characteristics as bar code, shape, and weight. The commands define the objects that exist and that should be transformed, and the objects to be produced as a result of this transformation;

- Can report that process steps have been executed by defining the objects that have been transformed and the objects that have been produced;

- Can identify objects and report their characteristics; and

- Can be commanded to stay within a certain domain. A robotic Automation Module may be commanded, for example, to stay within a geographical envelope.

Interaction Primitive	Arguments	Explanation
Processing step command	Objects Objects	Command to take certain *objects* and transform them into other *objects*.
Processing step progress	Objects Objects	Progress report: the *objects* that have been processed and the *objects* that have been produced.
Identification	Object	Identification of an *object*.
Domain restriction	Domain	Command for an Automation Module to stay within a *domain*.

Table 6.1: Workstation Controller–Automation Module Interactions

Chapter 7

Decomposition of an Automation Module

We described the interactions of Automation Modules in Section 6.2. Here, we analyse these to determine which tasks an Automation Module should execute.

7.1   Analysis of Automation Module Interactions

A Workstation Controller commands Automation Modules to execute processing steps on objects. An Automation Module may be able to execute several, technologically similar, processing steps on various kinds of objects.

We have been referring to pieces of material with certain physical characteristics, some of which are to be modified as a result of processing steps, as 'objects', We continue to refer to them as 'objects', as distinct from 'obstacles' or 'effectors'.

'Obstacles' are static or mobile pieces of material in the domain of an Automation Module that cannot be controlled by the Automation Module.

'Effectors' are pieces of material that are controlled by an Automation Module to interact with the objects to be processed.

Figure 7.1 illustrates a typical setting with objects, obstacles, and effectors. Effectors may exist to manipulate, polish, spray, heat objects, etc. Examples of effectors are the tools of a machine or the parts of a robot arm connected by joints.

A 'tool' is an effector that is needed by an Automation Module to process specific objects and that may need to be replaced when other objects are to be processed.

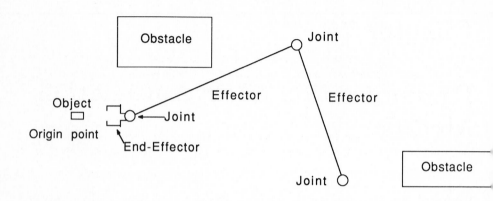

The object is to be moved from origin to destination.

Figure 7.1: Objects, Obstacles, And Effectors

An 'end-effector' is an effector that is intended to be in immediate physical contact with an object to be processed.

An Automation Module can execute processing steps on objects by coordinating its effectors so that they physically impact the objects to transform them. The effectors press the object, or heat it, or polish it, etc. The Automation Module must know which impact is required to process the objects as required. The required impact depends on the physical characteristics of objects and effectors.

In addition to *determining* the required impact of its effectors on objects, the Automation Module must know how to *realise* the required impact by steering its effectors while ensuring that the objects and effectors do not improperly interfere with obstacles. It should also be able to recognise objects, obstacles, and its effectors. Let us now elaborate on the steering of effectors and the recognition of objects.

Steering of Effectors. Let us consider a robotic Automation Module as in Figure 7.1 as an example to determine how an Automation Module in general can steer its effectors.

To realise a required impact of its effectors in order to transform an object, the Automation Module must know:

- the physical characteristics of the effectors, such as their mass, length, stiffness, and the relation between the position of an end-effector and a joint angle; and

- the actual state of the effectors.

Typically, most of the physical characteristics of an effector do not change during the execution of a processing step on an object. Think of the mass and stiffness of a robot arm. The state of an effector therefore depends on the few characteristics that can be changed. We refer to these as 'physical joints'. We use 'physical joint' as a generalisation of a robot's joint, the physical connection between two rigid effectors of which the angle and torque determine the position and forces on the robot's end-effector. Physical joints may characterise a position, temperature, or a pressure, etc. We will use 'joint variable' to denote a variable that describes the changeable characteristics of a physical joint.

A 'joint variable' describes a relation between two or more effectors, which is independent of relations with other effectors.

We say that a physical joint 'follows' a joint variable if the physical joint is modified so that the joint variable correctly describes the physical joint.

An Automation Module can control the impact of effectors on objects by steering physical joints, or, in other words, by determining and setting the required values of joint variables and ensuring that the physical joints follow the set points of joint variables.

Recognition of Objects.   An Automation Module should also be able to recognise objects and obstacles. To that end, it should be able to observe the physical world. An Automation Module may recognise the object depicted in Figure 7.1, for example, if:

- data describing the colour of the physical world shows a rectangular pattern with a colour that differs from its background and if the rectangular shape is characteristic for the object; or

- data describing an object with a certain weight and if the weight is characteristic for the object.

We conclude that an Automation Module needs data, describing portions of the physical world, to recognise certain objects and obstacles and to know the state of its physical joints. It may even have a need to recognise its own effectors, although sensed joint variables that reflect the physical joints will typically suffice. We refer to the data that describe the physical world as 'sensory data'.

We have discussed that an Automation Module should determine the required state of the physical joints and ensure that they reach that state, and obtain and interpret sensory data to recognise objects and obstacles. To determine the state of a physical joint, it is important to know whether the required state can be achieved but it is not important to know how the physical joint is actually modified. Similarly, to interpret the sensory data, it is not important to know how the data is obtained.

We therefore identify the following, relatively independent, globally defined Automation Module tasks:

- **Recognising and Modifying Objects**: interpret sensory data to recognise objects, effectors, and obstacles, and determine the required values of joint variables so that physical joints, which follow these variables, force effectors to physically interact with objects to execute the required processing steps on objects.

- **Providing Sensory Data and Steering Joints**: obtain sensory data and ensure that physical joints follow the joint variables.

The tasks are essentially different in purpose. The recognition and modification task deals with objects, which are described by multiple variables and their relations. The sensing and acting task deals with unrelated variables and serves to provide a direct and accurate relation between sensory data and joint variables and the corresponding parameters of the physical world. Figure 7.2 illustrates the different kinds of knowledge required to execute the Automation Module tasks.

7.2   Automation Module Controller and Devices

In compliance with the development strategy described in Section 3.1, we assign the relatively independent Automation Module tasks to distinct components. An 'Automation Module Controller' has to recognise and modify objects, and a 'Device' has to provide sensory data and steer the physical joints. Figure 7.3 illustrates our proposal for the organisation of an Automation Module.

Single Automation Module Controller.   An Automation Module Controller has to execute processing steps on objects by determining the required values of the

---

Recognition and Modification of Objects

Knowledge:

- how to modify the physical characteristics of objects by controlling the impact of effectors on objects;

- of the relation of joint variables and the impact of effectors on objects; and

- to interpret sensory data, which reflect the physical world, to recognise objects.

Providing Sensory Data and Steering Joints

Knowledge:

- to steer effectors so that their physical joints variables follow joint variables; and

- to provide sensory data that reflect the physical world.

Figure 7.2: Knowledge Required For Automation Module Tasks

126

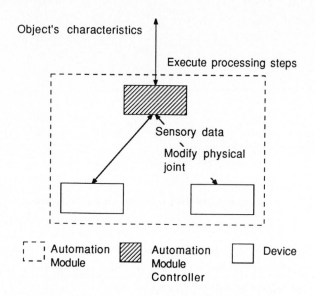

Object's characteristics

Execute processing steps

Sensory data

Modify physical
joint

Automation
Module

Automation
Module
Controller

Device

Figure 7.3: Decomposition Of Automation Module Into An Automation Module Controller and Devices

physical joints of its effectors [1]. It commands Devices to modify the physical joints according to certain joint variables. Some Devices provide the Automation Modules with the values of joint variables or with other sensory data. An Automation Module Controller may have to determine the required values of several joints, which may have to change simultaneously. Think, for example, of a situation where two effectors have to exert a force on an object simultaneously. The overview of all objects, obstacles, and devices can best be maintained by a single Automation Module Controller. We therefore propose that an Automation Module has only one Automation Module Controller.

Multiple Devices. The required values of the variables of the different joints as determined by the Automation Module Controller, can be realised independently of each other. Similarly, the different types of sensory data can be obtained independently. An Automation Module may therefore have multiple Devices.

---

[1] Please note the difference between 'Devices' and 'effectors'. A Device is a component on an Automation Module that can be requested to modify physical joints so that effectors are moved to impact on objects to be processed.

The following Sections describe some practical situations and discusses the role of an Automation Module Controller and Devices in those situations.

7.3 Modeling Real-Life Situations

We review some of the characteristics of robotic Automation Module Controllers, which have to displace objects, and of Automation Modules that should recognise objects in highly unstructured environments [58].

Robotic Automation Module Controller. A robotic Automation Module Controller may have to calculate joint angles and velocities to achieve a given position of its end-effector while avoiding joint motion limits, singularities, etc. The Automation Module Controller may try to optimise the trajectories of physical joints, for example, to find the shortest or fastest path of the end-effector towards the object [41]. It should select joint trajectories which ensure that several objects, some of which may be moving, do not improperly interfere. It has to apply 'reflex control' to avoid collision with suddenly appearing objects. There may be more joints that can be controlled than are strictly necessary to achieve a position: there are redundant degrees of freedom. Redundant degrees of freedom enlarge the possible number of paths to a destination which may be exploited to improve performance.

An Automation Module Controller which commands Devices to modify the positions or velocities of physical joints expresses the required modifications in joint variables, or, to use the common terminology, in joint coordinates. Commands from the Workstation Controller to the Automation Module Controller, however, are likely to be expressed in other coordinates, since the Workstation Controller does not know the degrees of freedom of Devices. This means that the Automation Module Controller has to perform coordinate transformations. It may also perform coordinate transformations because some calculations become simpler if they are based on specific coordinate systems. In case of redundant manipulators, joint coordinates systems become mandatory as it is in joint space that the redundancy is represented [111].

Also, an Automation Module Controller needs a common coordinate system if it has to coordinate various manipulators. Transformations are therefore needed between the several joint coordinate systems and the common system.

Depending on the situation, the Automation Module Controller may express required modifications in terms of position, or force, or both. It may prescribe the modifications accurately, or indicate boundary values. Reference [111] describes several strategies for hybrid motion and force control.

An Automation Module Controller assumes that the physical joints accurately follow the joint variables and that the sensory data accurately reflect the physical environment. In practice, the accuracy of joint variables and sensory data with respect to the physical world is limited so that calibration is required. An Automation Module Controller compares the joint variables and sensory data with some other sensory data that are supposed to accurately describe the physical world and that have a known relation with the joint variables and the other sensory data, The Automation Module Controller therefore knows the actual value of joints and sensory data as compared with the values provided by its Devices. It uses its knowledge of any difference to correct its interpretation of sensed joint values and sensory data.

Alternative Decompositions.   References [5,7] propose a decomposition of robotic Automation Modules that differs from ours. The Automation Module has to be able to process objects, like in our case. However, the equivalent of our Device, would not control the position of joints but of the end-effector. The control of joints is realised at lower levels in the control hierarchy.

We disagree with this decomposition because one should not separate the determination of the end-effector trajectory from the determination of the joint trajectories. Assume, for example, that a trajectory of the end-effector depicted in Figure 7.1, is considered that would follow an arc for 270 degrees in the counter clock-wise direction to move an object from origin to destination. One cannot ignore the movements of the effectors and joints that steer the end-effector. The obstacle in the lower right corner of Figure 7.1 would prohibit their counter clock-wise trajectories. In general, all joints, effectors, obstacles, and objects should be considered to determine the required joint variables.

Sensing.   We also want to expand briefly on the interpretation by an Automation Module Controller of descriptions of the physical world from Devices to recognise objects. An Automation Module Controller could recognise objects, for example, by model-based strategies: an object is recognised if a sufficient amount of sufficiently prominent features matches those of a stored model of the object. Different object representations and different matching strategies can be applied. A prominent technique to speed up the search is prediction. Once a feature has been identified, use its description and a priori knowledge to predict a next feature (for example, location). General problems are absence of sufficiently accurate description of features due to occlusion or noisy date and too large search spaces.

Another problem is perspective transformation. Suppose that an Automation Module Controller has to identify the geometry and location of a three dimensional object.

It may have to combine several sensory data which provide a partial description of the object. Each of these descriptions may have its own scale and geometrical orientation. These scale factors and geometrical orientation may depend on the value of the joints that determine, for example, focus depth, location, and orientation of a camera. The Automation Module should take them into account in order to determine the geometry of the object.

An Automation Module Controller will use sensory data to determine how it should modify joints to process an object. However, it may also be able to modify joints that affect a Device's capability to provide certain sensory data; for example, a camera on a robot arm. The modification of joints now serves to facilitate the collection of sensory data.

7.4   Automation Module–Device Interactions

Table 7.1 contains interaction primitives [2] to model a Device that:

- Can be commanded to realise certain joint variables. An Automation Module Controller may determine multiple joint variables pertaining to one physical joint, for example, the angle and torque of a robot joint [3].

- Can give sensory data.

Interaction Primitive	Arguments	Explanation
Set-Joint-Command	Joints	Required value of variables that describe a *joint*.
Sensory-Data	Sensory-Data	Values and types of *sensory data*.

Table 7.1: Interactions Of Automation Module Controller And Device

---

[2] Appendix A explains the purpose and meaning of interaction primitives.

[3] Practical commands to set joint variables [75] often define boundary values of joints, define joints as function of time, define joints in several coordinate frames, define joint values absolutely or relatively, etc. All these types of definitions can be modeled by constraints on the interactions, as explained in Appendix A.

Chapter 8

Decomposition of a Device

We described the interactions of Devices in Section 7.2. Here, we analyse these interactions to determine which tasks a Device should execute. We will keep this chapter relatively brief since it deals with sensing and control domains that have been discussed extensively in literature.

8.1   Analysis of Device Interactions

An Automation Module Controller commands a Device to change physical joints so that they reach the state described by the corresponding joint variables. Devices execute such commands and can provide Automation Module Controllers with sensory data that describe physical parameters of the environment.

Physical joint variables can be modified by controlling 'Actuators' as motors and heating elements that are connected to the physical joints. The impact of the Actuators on the physical joints depends on the (low energy) control signals they receive. A Device can therefore modify physical joints by issuing certain control signals to Actuators that are connected to these joints.

It can more accurately control the physical joints if it receives sensory signals that provide feedback about their actual state. It compares the actual state with the required state, and determines which control signals to issue so that any difference of actual and required state be reduced.

The sensory signals are provided by 'Sensors'. Sensors maintain a relation between some physical parameters and (low energy) sensory signals.

We may conclude that a Device can modify physical joints by issuing control signals to Actuators and interpreting sensory signals from Sensors. It has to determine which control signals it should issue so that the physical joints reach the state described by the joint variables given by the Automation Module Controller.

A Device also needs sensory signals from Sensors to provide sensory data to Automation Module Controllers. It has to process the sensory signals to extract the relevant data, to enhance the quality by reduction of noise, etc.

> **Example.** A camera is a Sensor, which issues sensory signals that contain data about the light emitted or reflected by objects in a scene. The sensory signals are to be decoded to extract information about the light intensity at various locations of the scene. Data from consecutive shots of the same, supposedly static, scene are compared to reduce noise. The resulting data are compared with a certain threshold value to determine whether data should be interpreted as black or white. Finally, a black and white pixel map is passed to an Automation Module Controller, possibly coded by descriptions of contours that separate the black and white areas.

We have identified different Device tasks, which are described below.

- **Servoing:** issue control signals to Actuators so that a physical joint follows the corresponding joint variables, and receive sensory signals from Sensors to observe the physical joints or to produce sensory data that describe physical parameters.

- **Sensing and Acting:** act on physical joints on the basis of control signals and provide sensory data on the basis of sensory signals.

These tasks are essentially different. The servoing task is a control task whereas the sensing and acting tasks maintain a relation between (control or sensory) signals and specific physical parameters.

The tasks are relatively independent as well. It is necessary to know the relation of control signals issued to an Actuator and its impact on the physical environment, but it is not necessary to know how the Actuator operates. Similarly, it is not necessary to know how a Sensor operates to interpret its sensory signals.

8.2 Device Controller and Sensors and Actuators

We decompose a Device into a 'Device Controller', which executes the servoing tasks, and a Sensor and Actuator, which execute the sensing and acting tasks respectively.

A Device Controller may need multiple Actuators or Sensors to execute its task; for example, a robot joint, which connects two effectors of a robot arm. The controller

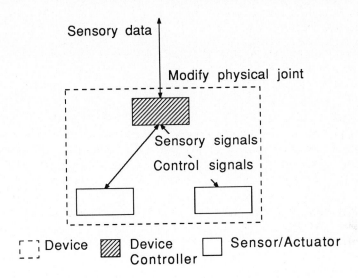

Figure 8.1: Decomposition Of Device Into Device Controller and Sensors and Actuators

may modify the joint by controlling two Actuators, which are connected to different effectors of the robot arm. It can also have redundant Actuators.

Similarly, a Device Controller can have multiple Sensors.

We therefore propose that a Device can have multiple Sensors and Actuators. Figure 8.1 illustrates the proposed organisation for a Device.

Chapter 9

Review of the Reference Model for MPCSs

We have developed a Reference Model for Manufacturing Planning and Control Systems. The model defines the global tasks of components of the total, more complex, MPCS and meanwhile allows one to keep sight of the relationship between the separate components. Hence, the model is an indispensable milestone in achieving integrated MPCSs.

The maxims that guided the decomposition process have led to important features of the model:

1. The separation of concerns maxim has led to MPCS controllers that are fairly independent. Although, the complexity of an individual controller may still be significant, their relative independence provides a way to deal with the overwhelming complexity of MPCSs: controllers can be considered independent of the internal behavior of other controllers.

2. The generality maxim has led to a Reference Model that despite its far-reaching consequences for the design of MPCSs at lower levels of abstraction can be usefully applied in the design of many MPCSs. Hence, it would save the top-level design efforts for those MPCSs.

3. The propriety maxim has led to a model that is efficient in the sense that it only describes tasks that are absolutely needed to run an MPCS.

Figure 9.1 illustrates the Reference Model.

136

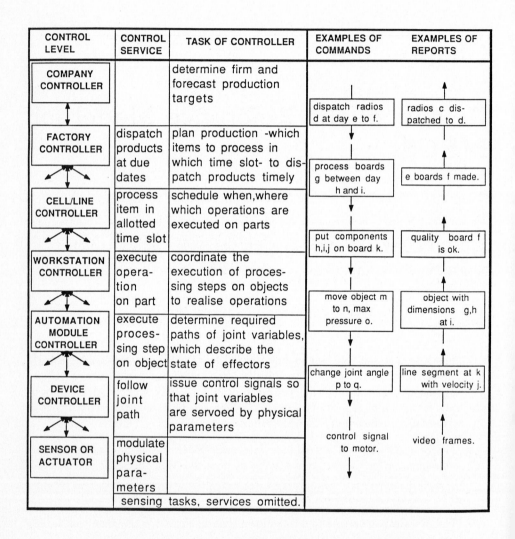

CONTROL LEVEL	CONTROL SERVICE	TASK OF CONTROLLER	EXAMPLES OF COMMANDS	EXAMPLES OF REPORTS
COMPANY CONTROLLER		determine firm and forecast production targets	dispatch radios d at day e to f.	radios c dispatched to d.
FACTORY CONTROLLER	dispatch products at due dates	plan production -which items to process in which time slot- to dispatch products timely	process boards g between day h and i.	e boards f made.
CELL/LINE CONTROLLER	process item in allotted time slot	schedule when,where which operations are executed on parts	put components h,i,j on board k.	quality board f is ok.
WORKSTATION CONTROLLER	execute operation on part	coordinate the execution of processing steps on objects to realise operations	move object m to n, max pressure o.	object with dimensions g,h at i.
AUTOMATION MODULE CONTROLLER	execute processing step on object	determine required paths of joint variables, which describe the state of effectors	change joint angle p to q.	line segment at k with velocity j.
DEVICE CONTROLLER	follow joint path	issue control signals so that joint variables are servoed by physical parameters	control signal to motor.	video frames.
SENSOR OR ACTUATOR	modulate physical parameters			
	sensing tasks, services omitted.			

Figure 9.1: **Reference Model Of MPCSs At A Glance**

Part III

Reference Model for MPCS Management

Chapter 10

Description of MPCS Management

In Part II of this work, we considered an inflexible MPCS, with a given product portfolio, production capacity, and production costs. In this Part [1], we consider a flexible MPCS with modifiable product portfolio, production capacity, and production costs.

In chapter 3, we introduced MPCS Management and MPCS Executor as the two building blocks of a flexible MPCS. We described MPCS Management in a general sense as a system that provides an MPCS Executor with application specific information to change the product portfolio, production capacity, or production costs. The MPCS Executor components, when loaded with the information from MPCS Management specific for a particular application, behave as the corresponding MPCS components in this particular application.

> **Example.** Figure 10.1 shows a 'Workstation Controller Executor', a component of the MPCS Executor, into which MPCS Management loads application-specific information so that it behaves as a Workstation Controller in a target application.
>
> One of the tasks of MPCS Management is to inform the Workstation Controller Executor which commands it may execute, and the addresses where it can locate the Cell/Line Controller Executor and the Automation Module Executors. MPCS Management also loads 'recipes', which tell how a specific operation should be executed: the order to assemble parts, the locations to store parts, etc. It will have to provide new recipes if the product portfolio changes.
>
> The Workstation Controller Executor interprets the information from MPCS Management, and can then execute the defined operations if required by the Cell/Line Controller Executor, send the corresponding status reports

---

[1]The chapters in Part III are based on Reference [16].

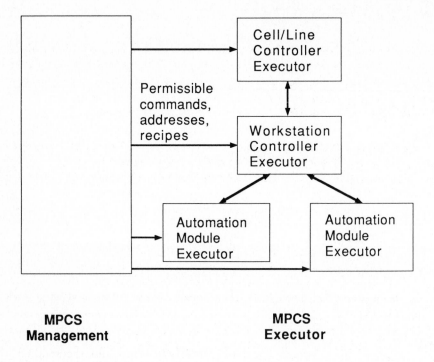

MPCS Management

MPCS Executor

Figure 10.1: MPCS Management Initiating A Workstation Controller Executor

to the addresses of the Cell/Line Controller Executor, command Automation Module Executors as prescribed by the recipes, accept and process status reports from Automation Module Executors, exchange parts with other Workstation Controller Executors, etc.

In chapter 3, we could not yet define which application specific information MPCS Management should be able to generate. However, having discussed the MPCS components in Part II, we can now define the application specific information more specifically. In this chapter, we therefore analyse the application specific information and describe the global tasks of MPCS Management in some detail as a prelude to a decomposition of MPCS Management in subsequent chapters.

In the following Section, we describe a flexible MPCS as a black box. Subsequently, we describe the decomposition of the black box into MPCS Management and MPCS Executor, and finally we describe the interactions of MPCS Management and its environment.

10.1 Flexible MPCSs

Figure 3.3 depicts a flexible MPCS as a black box. Like the inflexible MPCS, depicted in Figure 3.1, it interacts with its environment to establish its firm or forecast production targets, reports its material requirements, reports about its actual production, accepts raw materials, and dispatches products. Unlike the inflexible MPCS, it also interacts with the Company Controller to change its application by changing its:

- product portfolio. The Company Controller can propose a certain product portfolio by specifying the products that the MPCS should be able to manufacture. It can also allocate budgets that allow the MPCS to develop the products as well as the capability to manufacture them. The MPCS can report its actual product portfolio as well as how much of the budgets it has used to modify the product portfolio;

- production capacity. The Company Controller can propose a certain production capacity and allocate budgets that allow the MPCS to realise the required modifications of the production capacity. The MPCS can report its actual production capacity, as well as how much of the budgets it has used to modify the production capacity; and

- production costs. The Company Controller can propose to change the production costs and allocate budgets that allow the MPCS to realise the required modifications. The MPCS can report its actual production costs, as well as how much of the budgets it has used to modify the production costs.

The Company Controller, when requesting the flexible MPCS to modify its product portfolio, will typically also indicate the required production capacity and production costs. The MPCS should try to add new product to its product portfolio so that they can be manufactured in the required volume and at the required costs.

As shown in Figure 3.3, the flexible MPCS interacts with Customers and Suppliers to exchange resources. These resources include raw materials and products but also general resources such as machines and tools needed by the MPCS to process the raw materials.

10.2 MPCS Management and Execution

Following the approach for the development of a flexible MPCS, outlined in Section 3.3, we propose to decompose the flexible MPCS into two distinct MPCS components, i.e. Management and MPCS Executor. Figure 10.2 illustrates this. This decomposition reflects the inherent parallelism of MPCS management and execution: while the MPCS Executor performs its tasks in the current application, MPCS Management plans how the MPCS Executor should operate in another application.

The structure in Figure 10.2 also allows MPCS Management to incrementally add and delete knowledge of the MPCS Executors, while the MPCS Executor can continue to operate. It is indeed required that the MPCS Executor continues to operate when the application changes only partly, for example, when some new products are added to the product portfolio or when some machines are serviced. It would be impracticable to disable the entire MPCS Executor, create all the application-specific information from scratch, load the information, and enable the MPCS Executors. Instead, it must be possible to add and delete pieces of application-specific information concurrently with the MPCS execution.

Decomposition of MPCS Execution.   Figure 10.3 shows the result of a next decomposition step, i.e. the decomposition of the black box MPCS Executor in a structure of MPCS Controller Executors with globally defined tasks.

An 'MPCS Controller Executor' is a component of an MPCS Executor, which, when loaded with application specific information, behaves as an MPCS controller in that application.

A 'Factory Controller Executor' is an MPCS Controller Executor that when loaded with application specific information, behaves as a Factory Controller in that application.

Similar definitions apply to the other MPCS Controller Executors: the Cell/Line Controller Executor, the Workstation Controller Executor, etc.

We choose to give the MPCS Executor a structure that reflects the structure of the Reference Model for MPCSs developed in Part II. We did so because MPCS Controller Executors, when loaded with application specific information, behave as the MPCS Controllers in a given application, and will therefore execute their relatively independent tasks as defined by the Reference Model for MPCSs. The structure of MPCS Executor Controllers therefore meets the requirement for a Reference Model, explained in Section 3.2.2, that its components be relatively independent. Figure 10.3 therefore illustrates a Reference Model for MPCS Executors.

Figure 10.3 still depicts MPCS Management as a black box. In the following Sections, we investigate how MPCS Management interacts with its environment. In subsequent chapters, we analyse these interactions to decompose the black box which, using the decomposition strategy explained in Section 3.4, will result in a Reference Model for MPCS Management.

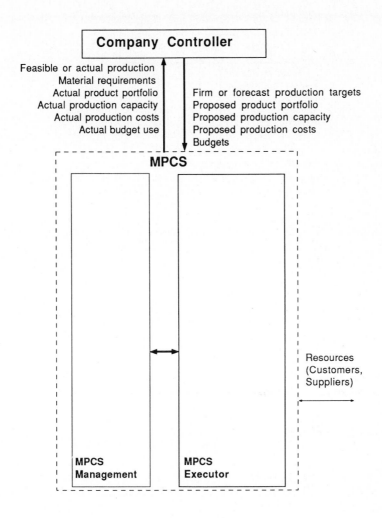

Figure 10.2: Flexible MPCS Consisting Of MPCS Management
And MPCS Executor

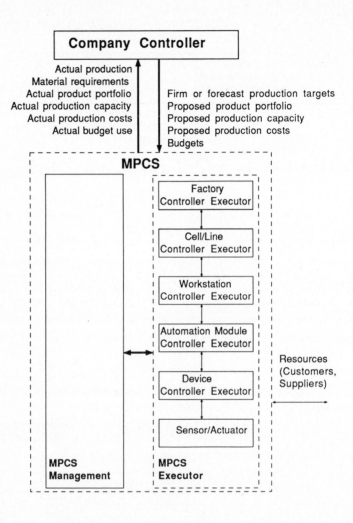

Figure 10.3: Decomposed Representation Of MPCS Executor

10.3 Interactions of MPCS Management and Its Environment

We first investigate the interactions of MPCS Management and the Company Controller. Subsequently, we investigate the interactions of MPCS Management and MPCS Executor.

10.3.1 Interactions with the Company Controller

We discussed in Section 10.1 that the Company Controller can request a flexible MPCS to modify its product portfolio, production capacity, or production costs.

MPCS Management has to investigate whether the MPCS Executor can be prepared for these changes. It has to consider, for example, the design of products and development of machines. We therefore propose that MPCS Management, as a component of the flexible MPCS, interacts with the Company Controller to establish whether product portfolio, production capacity, and production costs can be changed. MPCS Management then also reports to the Company Controller about the budgets needed and used to realise these changes, as well as the actual product portfolio, production capacity, and production costs.

The Company Controller determines the actual production targets for its flexible MPCSs. MPCS Management needs to know these targets to prepare the MPCS Executor for the realisation of the targets and to determine whether the MPCS Executor is performing well enough. We therefore propose that MPCS Management interacts with the Company Controller to establish forecast and firm production targets. MPCS Management also provides the Company Controller with feedback about the actual production.

MPCS Management prepares the MPCS Executor so that it can pursue the production targets and passes these targets to the MPCS Executor.

We have described the interactions of MPCS Management and the Company Controller. We will now investigate the interactions of MPCS Management and the MPCS Executor.

10.3.2 Interactions with the MPCS Executor

When MPCS Management has prepared the MPCS Executor for operation in a certain application, it can forward the firm or forecast production targets from the Company Controller to the MPCS Executor. The MPCS Executor can respond with reporting its material requirements and the status of the actual production to MPCS Management.

Note that the reports about the actual production may indicate that the required materials have not yet arrived. In such cases, MPCS Management may decide to use alternative materials for the production and generate new material requirements for the Company Controller. In other cases, MPCS Management will simply forward the material requirements and reports about the actual production to the Company Controller.

Let us now focus on the interactions of MPCS Management and MPCS Executor to prepare the latter for operation in a new application defined by new product portfolio, production capacity, or production costs.

Recall that the MPCS Controller Executors, when loaded with information specific for a particular application, operate as the corresponding MPCS Controllers in that particular application. The MPCS Controller Executors should therefore execute the tasks as defined by the Reference Model for MPCSs (see part II), while also adhering to some constraints defined by the application specific information from MPCS Management. These constraints should not violate those definitions of the Reference Model in Part II that are independent of the application, but they may:

- constrain aspects of the tasks that the Reference Model in Part II, specifies only non-deterministically; and

- change the values of constraints that, according to the Reference Model in Part II, are explicitly assumed to be stable in a given application.

Let us therefore consider the Reference Model for MPCSs, discussed in Part II, to investigate which aspects MPCS Management may specify deterministically and which constraints it may change.

Specifying Constraints Deterministically. The Reference Model for MPCSs specifies two kinds of constraints non-deterministically, i.e. the actual configuration of an MPCS and the procedures according to which an MPCS Controller executes its task. MPCS Management may therefore specify these constraints deterministically for the MPCS Executor so as to modify the configuration of the MPCS Executor and to define the procedures as to how the MPCS Controller Executors should execute their tasks.

A *'complete configuration'* of an MPCS Executor defines for all its MPCS Controller Executors and Sensors and Actuators with which other MPCS Controller Executors or Sensors or Actuators each of them may interact.

A *'complete configuration description'* defines for an MPCS Controller Executor with which MPCS Controller Executors and Sensors and Actuators it may interact, their identities, and their controller type.

MPCS Management could change the complete configuration of an MPCS Executor by generating a 'complete configuration description' for each of its MPCS Controller Executors and passing these to the MPCS Controller Executor involved. Further, MPCS Management could have to provide the MPCS Executor with physical MPCS Controller Executors, Sensors, Actuators, and effectors.

We call the prescription as to how an MPCS Controller should execute its task a 'planning and control procedure', or 'control procedure' for short.

A *'control procedure'* prescribes how an MPCS Controller Executor should react to commands and status reports so that its behaviour is optimised with respect to certain performance measures without violating the rules that prescribe how it should cooperate with other MPCS Controller Executors (its external functionality).

MPCS Management could generate control procedures and load these into the MPCS Controller Executors to improve their performance.

We conclude that the control procedures and complete configuration description are potential elements of the application specific information for the MPCS Executor that may be generated by MPCS Management.

Changing Application Specific Constraints. The constraints that, according to the Reference Model for MPCSs, are assumed to be stable in a given application, and that consequently may be modified by MPCS Management to change the application, have been discussed throughout the development of the Reference Model in Part II. We briefly summarise them here in the form of definitions.

The *'Topology of inventory points'* lists the jobs to be executed to manufacture a specific product in terms of the items they consume and produce, and their precedence relations.

Figure 4.4 depicts a topology of inventory points.

148

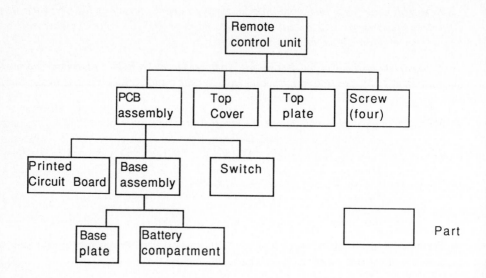

Figure 10.4: A Bill Of Material

A 'Cell/Line Executor allocation' identifies the Cell/Line Executors that may be ordered to execute jobs mentioned in a topology of inventory points, and their estimated job lead times.

Figure 4.7 depicts a Cell/Line Controller allocation.

A 'Bill Of Material', or 'BOM' lists the operations to be executed to realise a job that manufactures specific items, in terms of the parts they consume and produce, and their precedence relations.

Figure 10.4 illustrates a BOM.

A 'Bill of Process' identifies the Workstation Executors that may be ordered to execute specific operations mentioned in a BOM and their estimated operation lead times.

A 'Recipe' lists the processing steps to be executed to realise a specific operation that manufactures specific parts, in terms of the objects they consume and produce, precedence relations, the allowed tolerances, and the circumstances under which they have to be executed.

An 'Automation Module Executor allocation' identifies the Automation Module Ex-

ecutors that may be ordered to execute processing steps mentioned in a recipe and the expected lead times.

'Object characteristics' define the physical characteristics of objects to be processed by an Automation Module Controller Executor, of the obstacles in its domain, and of the effectors it uses to process the objects.

'Device characteristics' define the physical characteristics of Sensors and Actuators to be controlled by a Device Controller Executor.

We refer to the constraints that apply to a Factory Controller Executor (which coordinates the execution of jobs), to a Cell/Line Controller Executor (which coordinates the execution of operations), and to a Workstation Controller Executor (which coordinates the execution of processing steps) as 'production parameters'.

'Production parameters' define the Topology of inventory points, Cell/Line Executor allocations, BOMs, BOPs, Recipes, and Automation Module Executor allocations for an MPCS Executor in a specific application.

The production parameters define for a given application which *services* an MPCS Controller Executor can invoke, and which other MPCS Controller Executors it can use as *resources* for the execution of those services. Table 10.1 lists the steps and resources of the production parameters. The fact that production parameters define steps is a result of our partitioning of the manufacturing of a product into the discrete steps 'job', 'operation', and 'processing step'.

The steps and resources in the production parameters are related but can be changed independently to a certain extent.

> **Example.** The Workstation Executors defined in a particular BOP must be able to execute the operations defined in the related BOM to produce a Printed Circuit Board. There are two identical groups of Workstation Executors that can execute the component placement, solder, inspect, and transport operations defined in the BOM. The BOP can be changed without changing the BOM because Workstation Executors in both groups can execute the same operations. Such changes in the BOP are used to allocate the production of PCBs to one of the groups depending on the overall load and utilisation.

We refer to the constraints that apply to the Automation Module Controller Executor and Device Controller Executor as 'machine parameters'.

'Machine parameters' define the Object characteristics and Device characteristics

MPCS Controller Executor Type	Production Parameter Type	
	Services	**Resources**
Factory Controller Executor	Topology of inventory points	Cell/Line Executor allocation
Cell/Line Controller Executor	Bill Of Material	Bill Of Process
Workstation Controller Executor	Recipe	Automation Module Executor allocation

Table 10.1: Production Parameters

for an MPCS Executor in a specific application.

The machine parameters do not have the distinction between steps and resources as the production parameters. Such a distinction is not well possible because the processing steps have a continuous character.

MPCS Management Interactions. We may conclude that the Reference Model for MPCSs leaves room for four kinds of application specific information that *may* be generated by MPCS Management, i.e.:

- complete configuration descriptions;
- control procedures;
- production parameters; and
- machine parameters.

As we will show below, all four types of application specific information mentioned above have to changed by Management to change the product portfolio, production capacity, or production costs.

A change in the product portfolio:

- may require changes in the complete configuration and therefore in the complete configuration description since extra MPCS Controller Executors, Sensors, or Actuators may be needed for the manufacturing of new products;

- may require changes in the control procedures to provide any new MPCS Controller Executor with new control procedures that are more suitable for the new product portfolio;

- will certainly require new production parameters since new jobs, operations, and processing steps will have to be defined to manufacture the new products; and

- may require some changes in the machine parameters if new or modified Automation Module Executors are required.

A change in the production capacity or production costs:

- may require changes in the complete configuration description. To increase the production capacity, it may be needed to add extra Sensors, Actuators, or MPCS Controller Executors. To reduce the production costs, it may be required to remove some of them. It may also be required to replace MPCS Executor components for urgent or preventive maintenance;

- may require changes in the control procedures since these have an immediate impact on the production capacity and production costs;

- may require changes in the production parameters since these can have a significant impact on the performance and efficiency of an MPCS Executor. A BOM, for example, may define operations that will appear to become bottleneck operations and degrade the performance of the MPCS. Similarly, a BOM defines the precedence relations of operations. It therefore determines whether some operations could be executed in parallel, which would reduce the throughput time;

- may require changes in the machine parameters since these affect the performance of Automation Module Executors.

Application-Specific Information.   We conclude that MPCS Management, in order to modify the MPCS application, should be able to modify the complete configuration of the MPCS Executor and provide the MPCS Executor with complete configuration descriptions, control procedures, production parameters, and machine parameters.

In addition to the application specific information, MPCS Management should also be able to add to or remove from the MPCS Executor physical MPCS Controller Executors, Sensors, Actuators, or Effectors.

We also require that MPCS Management be able to measure its effects on the performance and efficiency of the MPCS Executor.

Figure 10.5 illustrates the interactions of MPCS Management and its environment.

Before concluding this chapter, we spend a few words on the relation of control procedures on the one hand and production and machine parameters on the other hand.

Production and Machine Parameters As Parameters of Control Procedures. Entirely new control procedures would be required for each new application if the control procedures could not be parameterised. Developing new control procedures for each new application would be a time consuming and expensive activity.

It makes sense, therefore, to develop control procedures that are parameterised so that changes of the parameters adapt MPCS Controller Executors to many new applications. Typically, the control procedures work satisfactorily for certain ranges of the parameters. The control procedures have to be replaced if the parameters are no longer within these ranges.

The production parameters and machine parameters can act as parameters for the control procedures. The following example illustrates that a BOM can be viewed as a parameter for scheduling procedures for a Cell/Line Controller.

> **Example.** A scheduling procedure allows a Cell/Line Controller Executor to schedule operations satisfactorily. These operations and their precedence relations are defined by BOMs. The scheduling procedure takes the list of jobs to be executed, and schedules the operations that, according to the BOM, have to be executed to realise the job, without violating the precedence constraints defined by the BOM.
>
> The Cell/Line Controller Executor can schedule operations for new items if it is loaded with BOMs for these items. However, the scheduling procedure works only satisfactorily for BOMs with strictly sequential precedence relations of the operations. The introduction of a BOM with parallel operations would require the introduction of a new scheduling procedure.

Nomenclature. For reasons of brevity, we henceforth refer to MPCS Executor components with the names we gave to the corresponding MPCS components. We use, for example, 'Workstation Controller' whereas 'Workstation Controller Executor' would be more accurate. Similarly, we speak of 'Controllers' instead of 'MPCS Controller Executors'.

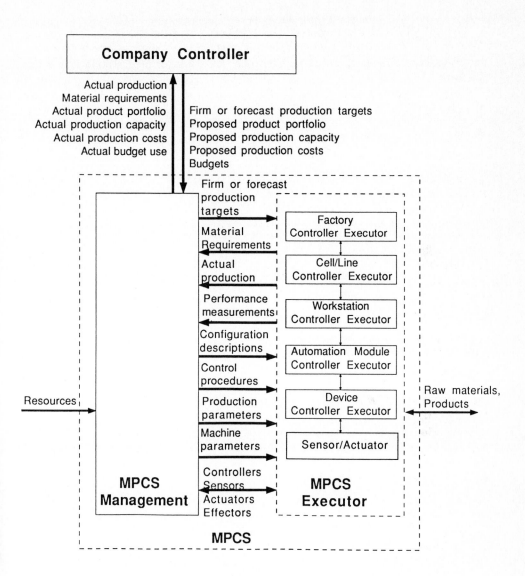

Figure 10.5: Interactions Of MPCS Management And Its Environment

154

We use 'MPCS Management' and 'Management' interchangeably. Further, we use 'Machine' for 'Automation Module Executor' and we use 'logistic controller' for Factory Controller Executor, Cell/Line Controller Executor, or Workstation Controller Executor.

Chapter 11

Decomposition of MPCS Management

In the previous chapter, we have described how Management interacts with the Company Controller and MPCS Executor. In this and subsequent chapters, we will decompose the black box Management into a structure of Management components. We will follow the decomposition strategy outlined in Section 3.3.4 to ensure that we identify Management components with relatively independent, though globally defined tasks. We call the result of this decomposition a 'Reference Model for MPCS Management'.

11.1 Analysis of MPCS Management Interactions

The Company Controller can request Management to change the product portfolio, production capacity, or production costs. Furthermore, it outlines production targets and allocates budgets. We will first investigate which tasks Management should execute to change the product portfolio. This will reveal how Management can, in principle, prepare an MPCS Executor to manufacture a given product. Once we know how a product can be manufactured in principle, we can probably determine as well how Management can decide that a product be manufactured differently so as to change the production capacity or production costs.

11.2 Changing a Product Portfolio

To prepare the MPCS Executor for the manufacturing of a new product, MPCS Management should, amongst others, determine how to:

- technologically transform raw materials into the required product;

- develop and install an MPCS Executor; and

- manufacture products, i.e. to determine which MPCS Executor can efficiently transform raw materials to manufacture new and existing products.

We will elaborate on these items in the following Sections.

11.2.1   Development of a Process Plan

Processing Steps and Machines.   Products result from the processing of material. For each product, Management must determine which processing steps, such as pressing, moving, heating, grinding, and blending, should be applied.

Management has to order processing steps in a technologically feasible manner. Suppose, for example, that two materials can be mixed only when they have reached a certain temperature. Management should then ensure that the materials are properly heated before they are mixed. Typically, many alternative sequences of processing steps are technologically feasible.

Each processing step is to be executed by a Machine (Automation Module Executor) [1]. Management therefore has to ensure that the MPCS Executor contains Machines that can execute the processing steps. Management may have to modify existing Machines, develop new ones, or select other processing steps to find processing steps that can be executed by Machines.

> **Examples.** A Machine can drill holes with various diameters. However, it cannot drill the hole required to make a specific product. The capability of the Machine is modified by giving it an additional drill with the required diameter.
>
> A Machine is developed to produce metal parts in large quantities. It is decided to design and build a new Machine because these parts do not resemble other parts and will be manufactured in sufficiently large quantities to justify the investment.
>
> It is too expensive to develop a Machine that can place components on a

---

[1]Recall that the Reference Model of MPCSs defines Automation Modules as the components that can execute processing steps.

Printed Circuit Board relatively close to each other. Instead, products will designed with components relatively far apart.

Process Plans.   Management should select processing steps to manufacture a product and ensure that these processing steps can be executed by Machines, which either exist or have to be developed. A proposal of the processing steps and Machines that could be used to execute them is expressed in a 'process plan'.

A 'process plan' lists processing steps, the conditions under which they have to be executed, their duration and technological order, to manufacture a specific product, and the Machines that can execute the processing steps or can be prepared to execute the processing steps, and the definition of the objects that should result from the execution of processing steps.

Figure 11.1 illustrates a process plan.

Impact of Process Plans on Performance.   Typically, Management can define numerous technologically feasible process plans but should select one or a few. The selection should be based on the manufacturing costs of a product and the required production volume. A process plan gives important indications about the costs and potential volume:

- A process plan specifies the raw materials needed to manufacture a product so that the material costs are known. It also specifies the Machine usage so that the processing costs are known. Further, a process plan affects various logistical costs such as the amount of tool transport required, the dependence on critical materials that may not always be available, the chance that certain machines are too heavily loaded and become bottlenecks, etc.

- A process plan has various effects on the performance of an MPCS Executor and the potential production volume: how much time various processing steps need, whether many processing steps can be executed in parallel, etc.

Process plans should satisfy 'operability constraints' to allow the efficient manufacturing of products in the required volumes.

'Operability constraints' are constraints on process plans that allow an MPCS Executor to efficiently manufacture products according to the process plan.

Table 11.1 lists some typical operability constraints on the reliability, lead times, frequency, quality, and allocation of processing steps in process plans.

158

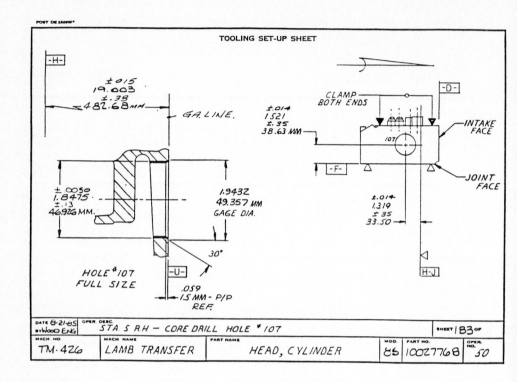

A processing step of a process plan, represented with drawings, which specifies the hole that should be drilled in part number 10027768 (cylinder head)
by Machine 'STA.5 R.H'.

Figure 11.1: A Sheet Of A Process Plan

Operability Constraint	Explanation
Process lead time	Limit on the minimum time in which a process plan can be executed.
Processing Step Lead Time	Limit on the duration of a processing step.
Variability	Limit on the variability in lead times of process steps.
Tool usage	Limits on tool changing and transportation, set-ups.
Machines	Preference to use certain Machines.
Similarity	Preference for process plans with many steps in common.
Quality	Minimum levels of quality, reliability, availability of processing steps.

Table 11.1: Typical Operability Constraints On Process Plans

Operability constraints can be used, for example, to discourage the usage of Machines that are far apart to avoid excessive transportation of parts. Operability constraints can also be used to favour process plans with a small amount of tool changes and set-ups [11,70].

We conclude that Management has to develop process plans and that meet operability constraints in order to assess whether it is technologically feasible to change in the product portfolio. Further, we conclude that the development of process plans and Machines are intimately linked. Process plans presuppose the capability to execute processing steps. Management therefore must ascertain that Machines, capable of executing the processing steps, are available or can be made available before selecting a process plan.

In the next Section, we will discuss the Management tasks to develop and install an MPCS Executor with a certain complete configuration and performance. Since we just concluded that the task to develop process plans includes the task to develop Machines, we will focus on the development of the remaining Controllers, the 'logistic Controllers' defined before.

11.2.2   Development of an MPCS Executor

Management has to be able to develop and install an MPCS Executor so that the Executor is operational at a required point in time. An MPCS Executor can be specified by defining its configuration and performance.

A 'configuration' defines for each of the logistic Controllers and Machines of an MPCS Executor with which logistic Controllers or Machines each of them may interact.

'Performance guidelines' define requirements for logistic Controllers regarding the lead times and delivery reliability with which they should execute orders to manufacture a product, execute a job, or execute an operation, and regarding the utilisation of the MPCS Executor components they coordinate.

Note that performance guidelines apply to logistic Controllers only. Performance requirements for Machines can be expressed in operability constraints.

'Performance indicators' describe the performance of logistic Controllers with respect to the requirements outlined by performance guidelines.

In Appendix B, we discuss performance indicators for the logistic Controllers.

Realising the Required Configuration.   Management can realise a required configuration by developing the required Machines and logistic Controllers and installing them. Moreover, it has to tell each logistic Controller with which other logistic Controllers and Machines it may interact. It should therefore give them a configuration description.

A 'configuration description' defines for a logistic Controller with which logistic Controller and Machines it may interact.

Realising the Required Performance.   Management can affect the performance of the MPCS Executor in several ways. It can develop Controllers that execute their control procedures faster. It can also improve the maintenance of the Controllers and Machines. Finally, it can also improve the control procedures.

Management can develop control procedures for the logistic Controllers on the basis of the performance guidelines, which define the required performance of a controller, and production parameters, which define which resources the Controller may use. A scheduling procedure for a Cell/Line Controller, for example, can be developed on the basis of the BOMs and BOPs for the Cell/Line Controller as well as the performance guidelines such as those that define:

- distribution and frequency of the jobs that will have to be executed;

- required job lead times;

- required percentage of the jobs that should be executed in time; and

- the expected operation lead time for certain Workstations.

Changing the control procedures will often be more effective than changing the logistic Controllers. However, there are cases where changing the logistic Controllers would give significant performance improvements as well. Take, for example, human logistic Controllers with non-automated processing of order files. The administrative processing of files can account for significant portions of the lead times so that automation of the administrative tasks of the Cell/Line Controller would result in significant performance improvements.

We have now discussed that Management should be able to develop process plans on the basis of product specifications and operability constraints and should be able to develop an MPCS Executor on the basis of a configuration definition, performance guidelines, and production parameters. In the following Section, we will focus on the selection of process plans and MPCS Executor.

11.2.3 Selection of Process Plans, MPCS Executor, and Production Parameters

Assuming that Management can develop process plans and MPCS Executors, it still has to decide which process plans to use and which MPCS Executor to develop so that the Executor can efficiently realise the production targets.

Management cannot choose process plans and MPCS Executor independently:

- The MPCS Executor should contain the Machines required to execute the process plans; and

- The specific allocation of the processing steps of a process plan to the Machines in a certain Executor configuration greatly affect the performance of the MPCS Executor.

There are two issues regarding the effects of processing step allocation on Executor performance. Firstly, the selection of processing steps and Machines, often referred to as 'machine balancing' or 'line balancing', affects the throughput and Machine utilisation.

Suppose that a number of objects have to visit several Machines to undergo a series of processing steps. Depending on how the processing steps and Machines have been chosen, some objects may need so much time at a particular Machine that they block objects that follow, which reduces the throughput. Or, objects may need so little time at a particular Machine that they leave it before objects that follow can visit it, which reduces the utilisation of the Machine. The selection of processing steps and Machines therefore affects throughput and Machine utilisation.

Secondly, the allocation of processing steps to Machines in a certain Executor configuration determines various characteristics of operations and jobs, which affect the Executor performance. Figure 11.2 can help to explain this. It illustrates that the processing step allocation determines:

– which processing steps are grouped into operations, which operations are grouped into jobs, which jobs should manufacture a product, and therefore affects the duration and temporal ordering of operations, jobs, and manufacturing of products. The chosen allocation may determine, for example, whether operations are defined that can be executed concurrently; and

– the allocation of operations to Workstations and of jobs to Cell/Lines, and therefore affects the utilisation of Workstations and Cell/Lines.

The processing steps could be allocated to the Machines in alternative ways if these yield alternative processing steps or if there are multiple Machines with similar capabilities. Management may decide to reject some of these alternatives.

Example. A process plan and configuration would yield two alternative Machines, which could each execute the same processing step. Management may decide to allocate the operation to one Machine exclusively if this would prevent the other Machine from being too heavily loaded.

In addition to allocating the processing steps to Machines, Management should ensure that the controllers of the Executor coordinate each other so that the right Machines are commanded to execute the allocated processing steps in the temporal ordering prescribed by the process plan. In other words, when the Factory Controller gets a request to manufacture a certain product, it should request the right Cell/Line Controllers to execute certain jobs at the right points in time. Similarly, the Cell/Line Controllers should request the right Workstations to command the right operations at the right points in time. And, similarly, the Workstations should request the right Automation Modules to execute the right Machines to execute the allocated processing steps at the right points in time.

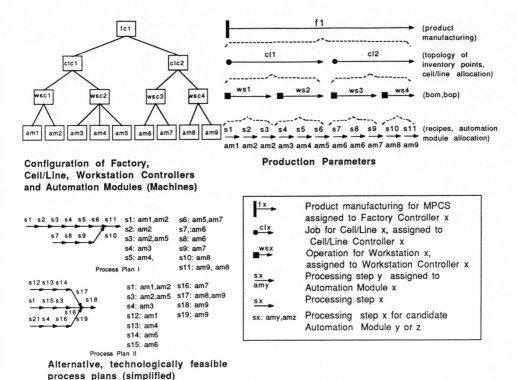

Configuration of Factory,
Cell/Line, Workstation Controllers
and Automation Modules (Machines)

Production Parameters

Alternative, technologically feasible
process plans (simplified)

The lower left corner shows alternative process plans. The upper right corner shows the operations and jobs for the MPCS Executor, shown in the upper left corner, to execute process plan I. According to process plan I, processing step 1 can be allocated to Machine am1 or am2. However, it has been allocated to am1. In this particular configuration, processing steps s1, s2, and s3 determine the net effect of operation ws1. Operations ws1 and ws2 determine the net effect of job cl1, etc.

Figure 11.2: Process Plan And Configuration Determine The
Feasible Production Parameters

Note that the services and resources of the *production parameters* discussed before define how the processing steps of a specific process plan are allocated to Machines and which Controllers cooperate to execute this process plan. Management should therefore select the allocation of processing steps to Machines and the production parameters in view of the above described performance issues.

Only those production parameters are feasible that:

- lead to the coordination of Machines that can execute the processing steps defined by this process plan; and

- define that certain logistic Controllers and Machines should cooperate without violating the required configuration.

We summarise the above by concluding that Management has to select process plans and MPCS Executors. Further, it has to allocate the processing steps to Machines and determine which Controllers will coordinate the execution of the processing steps. The latter is expressed in the production parameters.

Selection of Process Plans, MPCS Executor, and Production Parameters. Process plans, configuration and performance guidelines of the MPCS Executor, and production parameters all affect the MPCS Executor performance. Management, therefore, has to select these in a coherent fashion. Typically, it has to select among alternative combinations of these.

Example. Management should increase the performance of the portion of the MPCS Executor illustrated by Figure 11.3.

Assume that Management can measure four performance indicators. A first one describes how well a Cell/Line meets its due date commitments (lead time reliability). A second one describes the Cell/Line's throughput, which is a measure of the number of items the Cell/Line can process in a given period of time. A third performance indicator describes the quality of the parts produced by the Workstations in the Cell/Line, and a fourth one describes the operation lead times of the Workstations.

Suppose that the Workstations produce parts of a good quality with acceptable operation lead times, whereas the the Cell/Line meets its due dates reliably but has a low throughput. Thus, the performance indicators help Management to identify opportunities for performance improvements measures with a high potential pay-off. In this case, they point to performance improvements of the Cell/Line Controller rather than of the Workstations. Further, they indicate that its throughput should be improved.

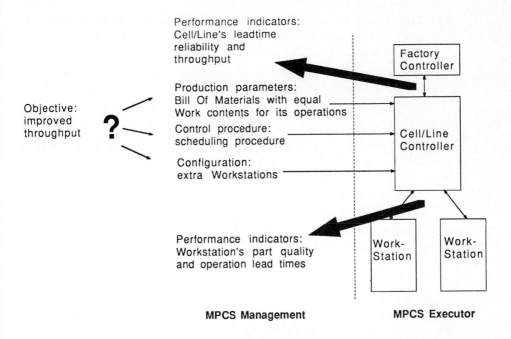

Figure 11.3: Performance Guidelines, Production Parameters, and Configuration Affect The Executor Performance

Management can try to reduce improve the throughput in three different ways:

1. Management can try to change the performance guidelines for the Cell/Line Controller, which could lead to revised scheduling procedures for the Cell/Line Controller;

2. Management can try to improve the performance by changing the production parameters. It could, for example, define a BOM with operations with equal work contents that can be scheduled efficiently;

3. Management could decide to change the configuration of the Cell/Line by adding Workstations, assuming that more Workstations could execute jobs faster.

Management will often search for process plans, configuration, performance guidelines, and production parameters in an iterative way. It takes a process plan and investigates whether it can change the MPCS configuration. It takes another process plan if it cannot find a suitable configuration, etc.

Often, Management will also use heuristic methods. It does not exhaustively explore all different combinations of process plans, configuration, performance guidelines, and production parameters. Instead, it decides heuristically, for example, not to consider changes in the process plans and to focus on changes in the configuration.

Example. An MPCS makes different models of TVs. The potential volume of a group of models can be assessed heuristically by considering the critical Machines used for all members of the group. The available capacity to insert components on a Printed Circuit Board may, for example, be the dominating factor to assess how many products can be made. The other Machines are typically not fully utilised and need not be considered to determine the potential volumes.

To prepare the MPCS Executor for the manufacturing of certain volumes of TVs, Management makes sure that sufficient insertion Machines are installed. Thus, it changes the configuration and production parameters; it does not change the process plans.

Development of Process Plans and MPCS Executor. While the process plans, operability constraints, configuration, performance guidelines, and production parameters cannot be *selected* independently, the process plans, MPCS Executor configuration, and control procedures can be *developed* independently. In fact, we already discussed in Section 11.2.1 that process plans can be developed on the basis of operability constraints and product specifications. And, we discussed in Section 11.2.2 that an MPCS Executor can be developed on the basis of a definition of its configuration, its performance guidelines, and the production parameters.

We have therefore identified the following relatively independent subtasks:

- Develop process plans on the basis of product specifications and operability constraints. The issue is how a product can be manufactured from a technological perspective.

- Develop an MPCS Executor on the basis of the required configuration and performance guidelines. The issue is how an MPCS Executor should be developed an kept operational.

- Select process plans, operability constraints, configuration, performance guidelines, and production parameters. The issue is to select MPCS Executor and process plans so that the MPCS Executor can efficiently make products by executing the technological processing steps in the process plans.

- Observe the Executor. The issue is to provide feed back information about its performance so that process plans, MPCS Executor, or production parameters can be modified to improve the performance.

We will refer to these subtasks as 'Product&Process Development', 'Supervision', 'Master Planning', and 'Monitoring' respectively:

- **Master Planning:** determine whether the MPCS application can be modified by:

  - selecting process plans and defining operability constraints for process plans;
  - selecting a configuration and defining performance guidelines for the MPCS Executor; and
  - selecting production parameters;

- **Product&Process Development:** develop process plans on the basis of product specifications and operability constraints, develop Machines required to execute the process plans, and define prescriptions for their maintenance;

- **Supervision:** modify and maintain the logistic Controllers, develop their control procedures, and install and maintain Machines on the basis of their maintenance prescriptions to realise the required Configuration and performance requirements; and

- **Monitoring:** provide information about the performance of MPCS Executor components.

Figure 11.4 shows that the distinct Management tasks require different kinds of knowledge.

Management will execute the above tasks to change the product portfolio. In the next Section, we will analyse whether it can also change the production capacity or production costs, or respond to production targets by executing these tasks.

11.3 Changing Production Capacity or Production Costs and Responding to Production Targets

To change the production capacity or production costs, Management would have to execute the tasks, discussed in the previous Section, that are also required to change the product portfolio. It may decide, for example, to change the configuration or the performance of Controllers to increase the production capacity or to reduce production costs. Similarly, it may decide that the process plans may have to be revised so as to meet new operability constraints. Similarly, it may have to change the production parameters, since these have a significant impact on the performance.

When Management receives production targets from the Company Controller, it will also have to select and install a suitable combination of configuration, process plans, performance guideline, and production parameters.

We conclude that Management can perform its task as described in Section 10.3 if it can execute the subtasks Master Planning, Product&Process Development, Supervision, and Monitoring as defined in Section 11.2.3.

Master Planning

Knowledge to:

- select production parameters, process plans, and configuration so that the application can be changed;
- determine operability constraints on process plans and performance guidelines for MPCS Executor controllers.

Product&Process Development

Knowledge of:

- product technology to design products that meet specifications;
- material technology to determine how materials can be processed to manufacture the required products;
- engineering disciplines and process technology to design and develop Machines;
- techniques to select process plans that meet operability constraints as much as possible.

Supervision

Knowledge of:

- planning techniques to modify the MPCS Executor while minimally disturbing the on-going production;
- engineering disciplines to develop and service logistic Controllers and to service Machines;
- mathematical techniques to develop control procedures; and
- control domains such as scheduling and inventory control.

Monitoring

Knowledge to:

- observe the MPCS Executor and analyse the observed data.

Figure 11.4: Knowledge Required For Different Management Tasks

11.4 Master Planner, Supervisor, Product&Process Developer, and Monitor

As Figure 11.5 illustrates, we propose to decompose Management into distinct components that execute the tasks described in Section 11.2.3. The:

- 'Master Planner' executes the Master Planning tasks;

- 'Product&Process Developer' executes the Product&Process Development tasks;

- 'Supervisor' executes the Supervision tasks; and

- 'Monitor' executes the Monitoring tasks.

Below, we briefly describe the interactions of these components. Key words referring to the interactions illustrated in Figure 11.5 are in italics.

Master Planner–Company Controller Interactions. The Master Planner will investigate the feasibility of changes of the *product portfolio*, *production capacity*, or *production costs* proposed by the Company Controller. It will also investigate whether the allotted *budgets* are sufficient to realise these changes.

It can request the Product&process Developer to develop *process plans* for this product by giving *product specifications* and *operability constraints*. The Master Planner can take the product specification from the description of the required *product portfolio* given by the Company Controller. The Master Planner will select operability constraints so that it is likely that the process plans allow the required *production capacity* and *production costs* to be realised.

It can define a *configuration* and *performance guidelines* for the Supervisor and request the Supervisor to install an MPCS Executor with the defined configuration and performance. It can also pass *production parameters* and a *shop floor calendar*.

A 'shop floor calendar' specifies when an MPCS Executor should be operational and when it is available for preventive maintenance.

It will consider the various *costs* reported by the Product&Process Developer and Supervisor and the *performance indicators* reported by the Supervisor to determine whether the modifications requested by the Company Controller are feasible.

If required by the Company Controller, the Master Planner will ensure that the feasible product portfolio, production volume, and production costs are implemented.

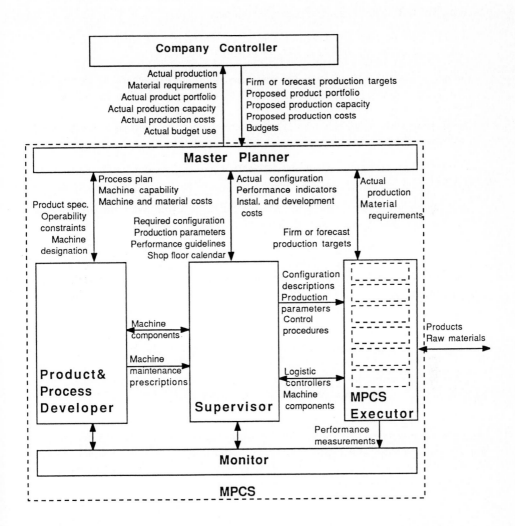

Figure 11.5: Decomposition Of MPCS Management

It selects a process plan. It also determines which Machines will be used to execute the processing steps in the process plan. It defines the *Machine designation* and reports the designation to the Product&Process Developer.

A 'Machine designation' defines the required capabilities of Machines in terms of the processing steps they should be able to execute.

On the basis of the Machine designation, the Product&Process Developer can prepare or develop Machines required to execute the process plan. The Master Planner will assume that these processing steps are executed in the amount of time specified by the process plan. Note that the Master Planner may require a certain availability and reliability level of Machines via operability constraints.

Similarly, Management defines the *production parameters* and passes these to the Supervisor.

It requests the Supervisor to install an MPCS Executor with a certain configuration, or requests it to modify the existing configuration.

Further, the Master Planner accepts *firm and forecast production targets* from the Company Controller that are feasible with respect to the product portfolio and production capacity. It will communicate these to the MPCS Executor once the Executor has been prepared to manufacture the products mentioned in the production targets.

Master Planner–Product&Process Developer Interactions. The Master Planner can ask the Product&Process Developer to develop a process plan for a particular product defined by a *product specification*. The process plan should adhere to certain *operability constraints* provided by the Master Planner.

The Product&Process Developer generates the required *process plans*. Note that these process plans describe the required machine capabilities but do not define which Machines should have these capabilities. The Master Planner tells the Product&Process Developer which process plans it will use and tells which Machines it has designated to execute certain processing steps of the process plan by reporting the *Machine designation*.

To realise the required Machine designation, the Product&Process Developer can develop any new Machines or modify existing Machines. It can receive existing *Machine components* from the Supervisor.

'Machine components' are Automation Module Controllers, Device Controllers, Sensors, Actuators, Effectors, Machine control procedures, or Machine parameters.

The Product&Process Developer passes all Machine components it develops or modifies to the Supervisor, which will install and service them.

The Product&Process Developer also passes *machine maintenance prescriptions* to the Supervisor to enable the Supervisor to service the Machines.

'Machine maintenance prescriptions' outline how and under which conditions a specific Machine should be serviced.

The Product&Process Developer provides the Master Planner with the *process plans* requested. Furthermore, it indicates the *costs* of materials and Machine usage for products manufactured according to the process plan. It also reports the *machine capability* to convey the availability of Machines to execute the processing steps.

'Material costs' are the costs of materials that are required to manufacture a specific product according to a certain process plan.

'Machining costs' are the costs of executing the processing steps to produce a certain product if these processing steps are executed by a certain Machine.

'Machine capability' defines the Machines that are available in terms of their capability to execute certain processing steps when requested to do so.

The Product&Process Developer monitors the Machines via the Monitor to check their actual performance.

Master Planner–Supervisor Interactions. The Master Planner can ask the Supervisor to install at a certain point in time an MPCS Executor with a *required configuration*. The Supervisor will install the required MPCS Executor. To that end, it can develop and install the *logistic Controllers*. It accepts *Machine components* from the Product&Process Developer and install these as well.

The Supervisor provides the logistic Controllers with *configuration descriptions* to inform them about the configuration.

The Supervisor will report to the Master Planner about the *actual configuration*.

The 'actual configuration' defines the configuration of the MPCS Executor and the Machines and logistic Controllers that are contained by the Supervisor.

Further,the Supervisor will report estimates of the *installation and development costs* of required modification of the configuration or performance.

'Installation and development costs' are the costs of developing, installing, and maintaining an MPCS Executor.

The Master Planner also passes *performance guidelines* to the Supervisor. The Supervisor will try to ensure that the logistic Controllers achieve the performance required by the performance guidelines. It can, for example, develop *control procedures* and load these into the logistic controllers. The Supervisor will monitor the logistic Controllers of the MPCS Executor via the Monitor and report about their performance to the Master Planner by means of *performance indicators*.

The Master Planner also passes a *shop floor calendar* to the Supervisor. The shop floor calendar allows the Supervisor to plan the installation of Machine components, logistic Controllers, and control procedures. Further, it allows the Supervisor to determine when preventive maintenance should take place.

The Master Planner also passes the *production parameters* to the MPCS Supervisor. The Supervisor needs these to develop suitable control procedures since, as we described in Section 10.3, only certain combinations of control procedures and production parameters work satisfactorily.

The Supervisor passes the production parameters to the logistic MPCS Controllers when they are needed to manufacture certain products and when the MPCS Executor has the required configuration.

Master Planner–MPCS Executor.   The Master Planner makes sure that the MPCS Executor has the configuration, performance, and *production parameters* to realise the required production targets, and forwards the *production targets* to the MPCS Executor.

The MPCS Executor reports to the Master Planner its *material requirements* and reports about the *progress* of the realisation of the production targets. The Master Planner forwards this information to the Company Controller.

In case of production targets from the Company Controller that call for products that are not manufactured yet, Management should determine the material requirements on the basis of process plans.

Monitor.   The Monitor can obtain performance data in various ways. It may monitor the performance continually, periodically, or ask components to be alerted if certain performance indicators pass certain thresholds, query the MPCS Executor components, etc.

It monitors the MPCS Executor on request from other Management components and may have to calculate averages, trends, correlations of various data, etc.

Typically, monitoring takes place to measure:

- Efficiency [56]. What is the status of a controller: up and processing orders, waiting for orders, waiting for materials, waiting for operators, waiting for planned or urgent maintenance, etc?

- Performance. How well do the Controllers execute their orders? Appendix B lists suggestions for performance indicators for the Controllers.

- Process control. Which events correlate with the product quality? The product quality may be insufficient despite flawless operation of Workstation Controllers and positive quality test results because some unknown events impact on the product quality. The correlation of such events and product quality needs to be determined.

Below, we discuss some specific tasks of Management components. These paragraphs can be skipped in a first reading.

Make or Buy Decisions.   The Master Planner can decide that certain objects, which would result from the partial execution of a process plan, are purchased as raw materials. In other words, a portion of a process plan will not be executed by the MPCS Executor, and the objects that would have resulted from the execution of this portion, are purchased from Suppliers.

The Master Planner can make such decisions, for example, on the basis of the material and machining costs and installation and development costs. These costs may justify that objects be purchased rather than manufactured.

Engineering Change Orders.   An improved product design, a change in the product specifications, a change in the layout or machine configuration, or recurrent flaws in the production are some of the events which may force changes in the on-going manufacturing of a product type. The request for such a change is referred to as the 'Engineering Change Order'.

The Master Planner must coordinate the updates of the production parameters, MPCS configuration, and process plans.

Capacity Expansion or Reduction.   The Company Controller may decide to allocate the future manufacturing of certain products to an MPCS, whereas the detailed product specifications are not known yet: "the production of radios will be allocated to MPCS X". The Master Planner has to change the capacity on the basis of incomplete process plans and refine the Configuration when product specifications and process plans are finalised.

Concurrent Design and Manufacture. In some types of industry, the manufacture of a product is already undertaken while the design of the product is still in progress. One starts to manufacture an airplane, for example, while some printed circuit boards that are part of the plane are still being designed. In such cases, the Master Planner releases production parameters with portions that have to be filled in later.

Our decomposition revealed to some extent how Management can do the tasks described in chapter 10. We further decompose the Product&Process Developer and Supervisor in chapters 12 and 13 respectively.

Chapter 12

Decomposition of a Product&Process Developer

We identified the Product&Process Developer in Section 11.4. In this chapter, we analyse its interactions to determine whether it can be decomposed further.

12.1 Analysis of Product&Process Developer Interactions

The task of the Product&Process Developer is to take a product specification and to develop a process plan for that product while meeting the operability constraints. Further, the Process&Product Developer can be required to modify or develop Machines so that these can execute certain processing steps.

Product Design. The Product&Process Developer needs knowledge of product technology to determine whether materials and components can be composed to make up a product. It needs to know, for example, how picture tubes, deflection units, tuners, amplifiers, and cabinets make up TVs.

A 'product design' defines a product's functionality, appearance, and composition of materials and components.

As follows form this definition, a product design does not prescribe how the MPCS Executor should manufacture the product. However, the design should define a product that can be manufactured. A design should not, for example, require that a screw be used to connect two parts if these parts would obstruct any tool that can be used

178

to insert the screw. Product designs should therefore meet 'manufacturability constraints'.

'Manufacturability constraints' are constrains on product designs so that the products can efficiently be manufactured given the available technological capabilities of Machines.

A few examples of manufacturability constraints follow:

Examples. The minimum distance of components on a printed circuit board so that they do not interfere with the placement device.

Preferred locations of components on a printed circuit board so that they can be placed more quickly by a specific component placement machine.

Preferred materials to limit the number of materials to be managed by the MPCS.

Product designs with as many similarities as possible to reduce the number of different parts.

Asymmetrical dimensions of parts to ease their feeding and recognition, if the parts need a certain orientation during an operation [89].

Machine Design. The Product&Process Developer should provide Machines that can execute processing steps to manufacture a product or parts of a product as specified by a product design. These Machines have to be developed or purchased if they are not yet available. A partial modification of a Machine's capability, for example by developing tools, may suffice to give it the required capabilities.

Process Planning. The Product&Process Developer has to decide in which steps a product with a certain product design is to be manufactured. Each step defines the shape, dimensions, colour of some materials at the beginning and end of the step. The steps, in effect, define the result of processing steps that need to be executed by Machines to manufacture the product.

It also has to be decided whether one or more Machines can execute the processing steps.

'Machine capabilities' define the processing steps that a certain Machine can possibly execute.

The capabilities of Machines and product design have to be conciliated. The Product&Process Developer should therefore define manufacturability constraints on the product design so that the product can be manufactured with the available or anticipated machine capabilities. It should also take the the operability constraints into account when defining the manufacturability constraints.

Machines and processing steps have to be selected out of the many possible alternatives. The processing steps, their temporal order, and the Machines that could execute the processing steps are all elements of a process plan. The process plan proposed to the Master Planner should be technologically feasible and meet the operability constraints as much as possible.

Conclusion. We identified the following, globally defined subtasks of Product&Process Development:

- **Product Design:** develop designs of products that meet the product specifications and manufacturability constraints;

- **Machine Design:** develop Machines that can execute certain processing steps; and

- **Process Planning:** consider Machine capabilities and operability constraints to define manufacturability constraints on product designs, and consider product designs, operability constraints, and Machine capabilities to develop process plans.

These tasks are relatively independent. Product Design can be done independently of Process Planning as long as the manufacturability constraints are met. Similarly, Process Planning can be done independently of Product Design, but should be based on the product design itself. Machine Design can be done independently of Process Planning and Product Design as soon as the processing steps have been specified. Product Design, Machine Design, and Process Planning also require different kinds of knowledge as Figure 12.1 illustrates.

12.2   Product Designer, Machine Designer, and Process Planner

As Figure 12.2 illustrates, we propose to decompose Product&Process Development into distinct components that can execute the tasks described in the previous Section. The:

- 'Product Designer' executes the Product Design tasks;

Product Design

Knowledge of:

- product technology, design tools, and design techniques;
- available components and materials and their characteristics that are relevant for product designs.

Process Planning

Knowledge of:

- material technology to determine how materials can be processed to manufacture the required products; and
- techniques to select process plans that satisfy the operability constraints as much as possible.

Machine Design

Knowledge of:

- engineering disciplines and process technology to design and build Machines and effectors, and define Machine maintenance prescriptions.

Figure 12.1: Knowledge Required For Different Product&Process Development Tasks

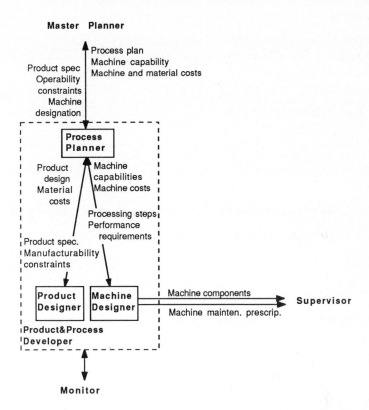

Figure 12.2: Decomposition Of A Product&Process Developer

- 'Machine Designer' executes the Machine Design tasks; and

- 'Process Planner' executes the Process Planning tasks.

We will explain the interactions of these components below. Key words characterising the interactions are in italics.

Process Planner Interactions.   The Process Planner accepts *product specifications* and *operability constraints* from the Master Planner. It has to return a *process plan* for the specified product.

To develop a process plan, it passes the *product specification* to the Product Designer. The Product Designer returns one or more *product designs*. Note that the Process Planner and Product Designer may pass product designs back and forth several times till a design has been produced that satisfies the Process Planner.

The Process Planner defines the processing steps that, if executed, will produce the product. The Process Planner negotiates with the Machine Designer whether the processing steps can be executed by Machines. The Machine Designer reports to the Process Planner the *feasible or actual Machine capabilities*.

The Process Planner searches for processing steps that realise the product design and that match the Machine capabilities. It translates the Machine capabilities and operability constraints into *manufacturability constraints* for the Product Designer.

The Process Planner passes technologically feasible process plan that can be executed by Machines to the Master Planner.

The Process Planner can also report the *machine and material costs* associated with a process plan. It calculates these costs on the basis of the *material costs* estimated by the Product Designer and the *machine costs* estimated by the Machine Designer.

The Process Planner may have to develop process plans for existing products if the Master Planner changes the operability constraints.

The Master Planner, having decided that a certain process plan will be used, outlines the *Machine designation* to the Process Planner. Thus, the Process Planner knows which Machines to prepare so that they can execute certain processing steps. It orders the Machine Designer to develop or modify the Machines so that they can indeed execute the processing steps.

Product Designer. The Product Designer should design products that meet the *product specification* and the *manufacturability constraints* given by the Process Planner.

It may not always be possible for the Product Designer to realise the product specification or to fully satisfy the manufacturability constraints. The product requirements may be too difficult to satisfy to begin with. Similarly, the design process may take so much time that it does no longer make sense to pass the design the Process Planner. In such situations, the Process Planner will renegotiate the requirements with the Master Planner.

The Product Designer may pass alternative product designs to the Process Planner. It may, for example, allow functionally equivalent components from different brands. This would allow the Process Planner to select the components that can be assembled most cost-effectively on the Machines it selects [48].

Machine Designer. The Machine Designer knows all Machines of the MPCS and can assess whether they can execute specific processing steps or can be modified to execute them. It will actually develop or modify Machines when requested to do so by the Process Planner.

It exchanges *Machines* with the Supervisor, which installs and services them. It also generates *Machine maintenance prescriptions* for the Supervisor.

Below, we discuss some specific tasks of the Product&Process Developer. These paragraphs can be skipped in a first reading.

Heuristic Development of Process Plans. Typically, process plans are developed using heuristics. Group technology is an example of a heuristic approach, essentially aimed at a reduction of the number of different parts and processes. It is a technique to rationalise part design and manufacturing, which also affects the creation of BOMs, and recipes.

The technique aims at exploiting similarity of parts and technological processes in design and manufacture [55,57]. The similarities are identified by means of elaborate coding and classification of the parts on the basis of their geometry, features, tolerances, materials, machining processes and requirements.

It can be investigated whether common parts can be used or designed if parts appear to be dissimilar but not entirely different. In the design process, the use of existing designs is encouraged.

Parts with similar shapes or features, or with similar processing requirements are grouped into families and allocated to a limited group of Automation Modules, Workstations, and Cell/Lines. The technique thus reduces the number of different parts. It also reduces the need for widely different processing capabilities and the need for the transportation of tools.

Group technology can also help in the development of recipes because the processing requirements of an operation, on a part, can be deduced from the coding and classification of the part.

Preliminary Designs. The Product Designer may make a preliminary investigation of whether a product can be designed. The preliminary design is passed to the Process Planner, which assesses whether it can make a process plan. The Product Designer can finalise the design if the initial evaluation of the design by the Process Planner is positive.

The Product Designer will also make preliminary designs if it is forced to do so by the requests of the Process Planner. Suppose, for example, that the Process Planner received a plan to double the production of portable radios, starting next year. The specific requirements of each type of radio are not known yet. However, it may be possible to develop preliminary designs that define such parts as tuner, amplifier, and wiring, without defining the overall product. These rough designs are passed to the Process Planner, which will, for example, negotiate with Machine Design the development of new Machines.

Chapter 13

Decomposition of a Supervisor

We described the global task of a Supervisor in Section 11.4. In this chapter, we therefore analyse its interactions to determine whether it can be decomposed further.

13.1 Analysis of Supervisor Interactions

The Supervisor should develop and maintain an MPCS Executor with a configuration and performance as outlined by the Master Planner. Furthermore, it should load the logistic Controllers with production parameters proposed by the Master Planner. Finally, it receives from the Master Planner the shop floor calendar outlining the time periods in which the MPCS Executor should be operational.

Configuration. The Supervisor can realise the required configuration by developing and installing logistic Controllers and by installing Machine components.

The Machine components are designed and built by the Product&Process Developer. The role of Supervision is confined to installation and maintenance according to the Machine maintenance prescriptions passed by the Product&Process Developer.

The Supervisor has to develop the logistic Controllers. It can use the Reference Model for MPCS Executors and develop complete functional specifications of logistic Controllers on the basis of the Reference Model. Further, it has to develop physical implementations of the Controllers.

Performance. The Supervisor can try to improve the performance of logistic

Controllers in three ways. It can try to develop *control procedures* that render a higher performance, it can develop logistic Controllers with a higher performance, or it can improve the availability of the logistic Controllers by improving their maintenance.

We discussed before that the Supervisor needs performance guidelines and production parameters to develop control procedures. Further, it needs to observe the logistic Controllers and Machines of the MPCS Executor to assess how the control procedures it has developed do perform.

Coordination. The installation of Controllers and loading of control procedures into these Controllers needs to be coordinated. Moreover, the installation of logistic Controllers has to be coordinated with the installation of Machines.

Other coordination issues involve the selection of performance improvement measures. Should, for example, maintenance be improved or the control procedures?

Conclusion. It appears from the above that the Supervisor should develop, install, and service various logistic Controllers and their control procedures, install and service Machines, develop and install control procedures, and coordinate all these activities.

The development of Controllers and control procedures cannot easily be done independently. The reason is that the control procedures have to be coded in a form that can be executed by the controllers. A control procedure may have to be coded, for example, in a natural language for a human scheduler but in a programming language for a computerised scheduling system. The development of control procedures and controllers is therefore related.

Maintenance and controller development are less dependent. One can service a controller on the basis of observations of its operation and their maintenance prescriptions. Maintenance descriptions need to be generated by the developer of the controllers.

We may conclude that the Supervisor execute the following, globally defined, relatively independent tasks:

- **Configuration Management:** realise required modifications of the MPCS Executor by coordinating the development, modification, and installation of of logistic Controllers including their control procedures, installation and maintenance of Machines, and installation of production parameters and configuration descriptions;

- **Logistic Controller Design:** develop logistic Controllers and their control procedures so that they meet their performance guidelines; and

- **Maintenance:** install and service logistic Controllers and install and service Machines.

Figure 13.1 illustrates the different kinds of knowledge required for the respective tasks.

13.2   Configuration Manager, Logistic Controller Designer, and Maintenance

We propose to decompose the Supervisor into components that execute the tasks defined in the previous Section:

- The 'Configuration Manager' executes the Configuration Management tasks;

- The 'Logistic Controller Designer' executes the Logistic Controller Design task; and

- 'Maintenance' executes the Maintenance tasks.

Figure 13.2 illustrates the proposed organisation of the Supervisor. We describe the interactions of the Supervisor components below. Key words characterising the interactions are in italics.

Configuration Manager.   The Configuration Manager receives from the Master Planner requests to install and maintain an MPCS Executor with a certain *required configuration*. Further, it receives a *shop floor calendar*, outlining the time periods in which the MPCS Executor should be operational, and it receives *performance guidelines* for the various logistic Controllers in the MPCS Executor. Finally, it receives *production parameters*, which, in fact, select a portion of the configuration that should manufacture a specific product.

The Configuration Manager tries to install an MPCS Executor as required by the Master Planner by requesting:

- Maintenance to install or remove logistic Controllers or Machines; and

Configuration Management

Knowledge of:

- techniques to plan modifications of the MPCS while minimising the impact on the on-going production.

Maintenance

Knowledge of:

- maintenance and installation techniques.

Logistic Controller Design

Knowledge of:

- specification and implementation techniques to build the logistic controllers.
- techniques to develop planning and control algorithms;
- control domains, such as scheduling and inventory control.

Figure 13.1: Knowledge Required For Different Supervision Tasks

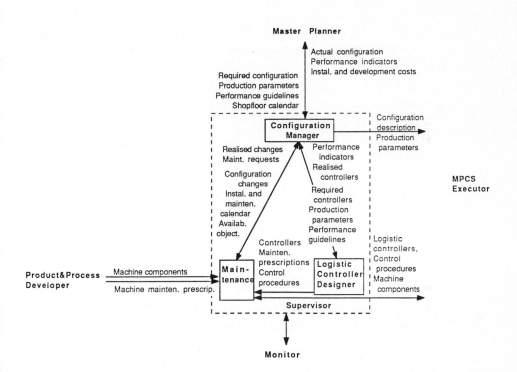

Figure 13.2: Decomposition Of A Supervisor

- the Logistic Controller Designer to develop logistic Controllers or to improve their performance.

The Configuration Manager can load *production parameters* and *configuration descriptions* into the logistic Controllers.

Maintenance.   The Configuration Manager requests Maintenance to realise *changes in the configuration.* Further, it passes an *installation and maintenance calendar.*

An *'installation and maintenance calendar' defines when which Machines or logistic Controllers have to be installed and when they may receive preventive maintenance.*

The Configuration Manager also passes *availability objectives.*

'Availability objectives' define the required availability of Machines and logistic Controllers.

Maintenance accepts *Machine components* to be installed from the Product&Process Developer along with *machine maintenance prescriptions.* Maintenance may also have an inventory of Machines that can be installed.

Maintenance can observe Machines in operation via Monitoring, and compare their performance with the conditions that, according to the machine maintenance prescriptions, indicate that preventive maintenance is required.

On the basis of these maintenance prescriptions, it can modify the machine components: the control procedures and machine parameters of the Machines as well as their Controllers, Sensors, Actuators, or effectors.

Similarly, Maintenance acquires *logistic Controllers* from the Logistic Controller Designer along with *maintenance prescriptions* and *control procedures* for the logistic Controllers. Maintenance may have logistic Controllers in storage that can be installed as well.

'Logistic Controller maintenance prescriptions' describe how and under which conditions a specific Logistic Controller should be serviced.

Maintenance installs the logistic Controllers of their control procedures if requested by the Configuration Manager. It will perform preventive maintenance, following the maintenance prescriptions, in time slots allowed by the installation and maintenance calendar.

Maintenance also performs urgent maintenance. It observes the MPCS Executor in operation, via the Monitor, to decide whether they behave in such a way that, according to the maintenance prescriptions, they need maintenance. If it decides that urgent maintenance is required, it will notify the Configuration Manager with a *maintenance request*. The Configuration Manager should authorise urgent maintenance.

Logistic Controller Designer. The Configuration Manager orders the Logistic Controller Designer to develop new logistic Controllers by specifying the *required controllers*. The Logistic Controller Designer develops these and passes them to Maintenance along with *logistic Controller maintenance prescriptions*. It reports to the Configuration Manager whether it has developed the required controllers by reporting which *logistic controllers* it has *realised*.

The Logistic Controller Designer should also develop control procedures that allow the logistic Controllers to meet the performance objectives as defined by the *performance guidelines*. The Logistic Controller Designer may also improve control procedures on its own initiative for example if it concludes that the performance guidelines are not met on the basis of feedback from the Monitor. It can report to the Configuration Manager that it wants to change a control procedure by issuing expected *performance indicators*. In all cases, the Configuration Manager determines when new control procedures are loaded into the Controllers and reports such with an installation calendar.

It is possible that the Logistic Controller Designer may not be able to develop on time the control procedures that meet the performance guidelines and instead may propose procedures that do not fully meet the objectives. This will ultimately force the Master Planner to search for further alternatives to improve the MPCS performance.

Below, we discuss some the Supervisor components in some specific situations. These paragraphs can be skipped in a first reading.

Positioning of Inventory. In Section 5.3, we discussed that Cell/Lines may contain some inventory to improve their performance. Specifically, we discussed the inventory in JIT systems and in buffers that compensate for unbalanced capacities. The required amount and allocation of this inventory depend on the layout, performance of Workstations, types of parts produced, and the required Cell/Line performance. For the sake of brevity, we refer to this type of inventory as 'catalyst stock'.

'Catalyst stock' is inventory of parts that allows a Cell/Line to improve its performance because and is replenished to a certain level independent of the execution of jobs by the Cell/Line Controller.

The Logistic Controller Designer can decide that catalyst stock is required to achieve a certain Cell/Line performance. If so, it has to develop two categories of procedures and load these into Controllers.

One category of procedures, if executed by various Controllers, will result in the creation of catalyst stock. More specifically, Cell/Line Controllers should take parts from inventory points and process them to create the required catalyst stock, and Factory Controllers should replenish the inventory points and report the material requirements.

Another category of procedures are required to maintain the catalyst stock. These are the scheduling and coordination procedures, which should enable the Cell/Line Controllers to realise the main performance objectives, but also to maintain catalyst stock at adequate levels. Examples of procedures that keep inventories at certain levels are "hedging point" strategies described in Reference [2].

Error Correction. During production, all kinds of unforeseen events may happen that cause failures. For example, a part gets blocked in a robotic Workstation. The robot arm has to withdraw, the part has to be dispatched and trashed.

The Logistic Controller Designer has to develop the procedures that allow the logistic Controllers to perform corrective actions. The Logistic Procedure Developer will be notified via the Monitor about the occurrence and nature of unforeseen events.

Chapter 14

Review of the Reference Model for MPCS Management

We have defined the role of MPCS Management and, using the strategy outlined in chapter 3, developed a structure of MPCS Management components with relatively independent, globally defined tasks.

The resulting Reference Model for MPCS Executor and MPCS Management can be used as a basis for the development of flexible MPCS Executor Controllers that can be tuned to operate efficiently in specific applications. It can also be used as a basis for the development of complete functional specifications of Management components.

The Reference Model now gives a complete picture of the activities that occur in a real production organisation. It decouples many computationally intensive tasks such as product design, process planning, development of control procedures, and many distinct planning and control tasks. It therefore allows us to understand these tasks and their relation with other tasks.

Figure 14.1 illustrates our proposal for an MPCS Management organisation. This Figure does not show the following details:

- the Machine Designer only needs information about the Machines.

- the Logistic Controller Designer consists of separate components responsible for the development of control procedures of Factory Controllers, Cell/Line Controllers, and Workstation Controller respectively.

- Several Management components can interact with Suppliers to accepts raw materials, Machines, tools, etc; and

194

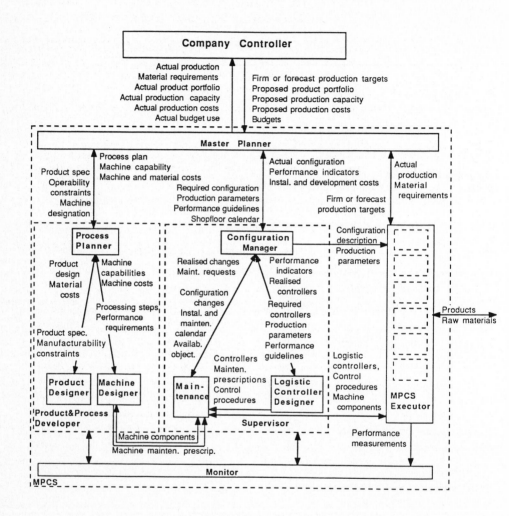

Figure 14.1: Reference Model For MPCS Management

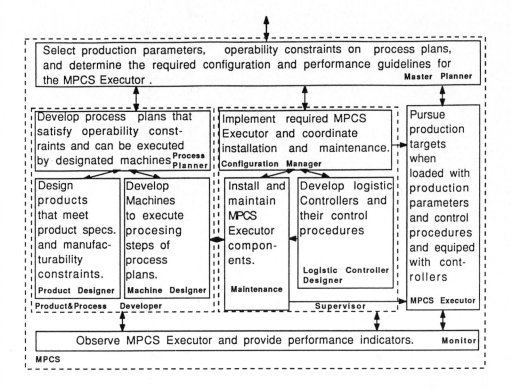

Figure 14.2: Tasks Of Management Components

Figure 14.2 summarises the task of the Management components illustrated in Figure 14.1.

It may be possible that the Management components can be decomposed further. Further decomposition of Management may reveal components that resemble MPCS Executor components. The Machine Designer, for example, may have warehouse components to store spare parts like the MPCS Executor stores parts of products. Similar Management and MPCS tasks can be assigned to the same components at a lower level of abstraction. That is why a real warehouse may contain parts for manufacturing and spare parts for Machine Design.

Other types of Management will arise when the physical implementation of the MPCS and its Management is considered. If we implement, for example, the MPCS with computers and networks, we will need management for them. Typically, we will also need a personnel department to recruit, train, and pay human operators that work in the MPCS.

Part IV

Conclusion

Chapter 15

Conclusions

15.1 Integrated Framework for Manufacturing Planning and Control

We substantiated a framework for 'integrated manufacturing control' by analysing which relatively simple planning and control tasks have to be executed to operate a Manufacturing Planning and Control System, which accepts production targets from a commercial department, and manufactures products accordingly. This framework defines objectives and constraints for control systems, be they human or automated, so that they contribute to the performance and flexibility of an MPCS in a manner that is visible for the commercial department.

The development and use of this framework implies a departure from common and conventional approaches that address planning and control problems in isolation, without validating the relevance of their objectives and constraints in the context of a larger system. These approaches run the risk of producing solutions that do not contribute to the overall performance of the entire system.

We started to describe an MPCS as a black box, precisely as it is seen by the commercial department. In several steps, we analysed how this black box could be realised by controllers that execute distinct MPCS subtasks. The result of this analysis is an organisation of MPCS controllers, which defines their structure, global tasks, and information requirements. The top-down analysis, starting from the black box MPCS to derive a structure of MPCS controllers, is in essence the approach we advocate to achieve integrated manufacturing control.

We also analysed and proposed an organisation for MPCS Management. Insight into MPCS Management is vital for flexible MPCSs because it helps to quickly introduce the manufacturing of new products, change the production capacity, and reduce the production costs.

We designed an MPCS and its Management on the basis of design heuristics that should ensure that they:

- can perform the function of the black box MPCS as described in Chapter 2;

- have the conceptual integrity, clarity, and modularity to allow a human to master and understand the system despite its tremendous, inherent complexity; and

- are generally applicable in the sense that they can, in principle, model the manufacturing of a wide variety of products, with a wide variety of planing and control procedures, with a wide variety of machines, in a wide variety of layouts.

The advantage, and uniqueness, of the ability to understand an MPCS should not be underestimated. It allows us to quickly develop MPCSs, to control them, to adapt them, to monitor them, etc. These are qualitative but nevertheless important characteristics of an MPCS in addition to the more traditional quantitative performance criteria like throughput.

'Separation of concerns', described in Section 3.4, is the most important design heuristic that we applied. To successfully apply this heuristic, *experience* and *vision* is needed in order to assess whether some concerns are significantly more related than others. The designed framework is not generally applicable if it can be shown that concerns were separated that in many applications are more closely related than concerns that were not separate.

15.2 Applications

The design of an 'idealised' MPCS, as expressed by the Reference Model can be used (and has been used) in at least three different way, which we briefly describe below.

1. Analysis and improvement of existing MPCSs.

   Existing factories manage to manufacture products. However, one could always question whether they can manufacture better, more, quicker, and cheaper.

   A Reference Model for an MPCS, being an idealised organisation for MPCSs, can be compared with existing MPCSs. Typically, it is required to 'translate' the generic Reference Model into a specific model of the local MPCS, expressed in the local terminology. While the Reference Model uses general terms like 'operation', the local model refers to specific operations like, for example, 'gun sealing' or 'applying graphite' [42,39].

The local model has to be mapped onto the existing organisation of people and systems engaged in planning and control tasks. The translations and mappings will only be partially successful since the existing MPCS and the idealised MPCS typically differ. It can now be decided whether or not the revealed differences should be eliminated. Note that the differences may as well serve to validate the Reference Model.

2. Development of control systems.

The Reference Model contains specifications of the tasks, constraints, and information requirements for MPCS controllers. These specifications can be viewed as top-level user requirements for physical implementations of those controllers. They are therefore the starting point for a systems engineering process to specify, design, and implement those controllers [20,28,21].

3. Development of planning and control algorithms.

The 'heart' of the controllers defined in the Reference Model is an algorithm that enables a controller to operate. The Reference Model and functional specifications of the inputs and outputs of the controllers can be viewed as objectives and boundary constraints for the algorithms.

Introduction Strategy. Almost invariably, people who want to apply theories of manufacturing planning and control systems have to deal with existing factories; fresh applications are quite scarce. Experiments are difficult since the on-going production should be disrupted as little as possible, if at all. It is therefore imperative that strategies be developed to migrate from an existing to an improved situation while disturbing the production minimally.

It may be difficult to convince people who run a factory of the benefits of an integrated approach. Like many researchers, they may not yet have a comprehensive overview of the entire manufacturing organisation as described by the Reference Model. The concept of an integrated MPCS may be difficult to understand because it requires abstraction skills and insights into a wide variety of subjects, which are traditionally taught in isolation. Inadequate understanding of the integrated MPCS will support the notion that myopic solutions are adequate. It may also lead to a perception of threat if the concept of integration is confused with total automation and consequent elimination of jobs and responsibilities or with systems that are too complex to be controlled.

In view of this, it might be a good idea to develop and introduce monitoring and performance analysis tools prior to introducing a, human or automated, integrated MPCS.

Such monitoring and performance analysis tools may assist people in understanding a factory as a whole and therefore may make them more receptive of integrated approaches and solutions. Moreover, these tools are less likely to be perceived as threatening because they do not deprive people of their decision-making authority, but rather support their decision-making.

These tools would allow various people in a factory to assess the performance of machines, departments, scheduling and planning systems, forecasting, etc. They should monitor such high level performance indicators as throughput, customer response times, and inventory turnover in addition to the more common performance indicators such as machine utilisation, repair times, etc.

An important issue regarding advanced performance analysis systems is which information they should collect and present. This is not a trivial matter and requires a deep insight into the performance indicators for the various planning and control tasks across all levels of the hierarchy, both in a qualitative and quantitative sense. Further, these systems will probably require advanced user interfaces to zoom in and out, and switch between monitoring on an aggregate and detailed level. Data collection and storage systems, which are available on the market, may serve as a platform.

The introduction of integrated systems into existing factories will often face the requirement that modifications of existing factories with be justified with respect to their return on investment. This also suggests that performance analysis merits our special attention. Many of today's performance indicators encourage sub optimising behaviour of people and systems rather than optimising the behaviour of the MPCS as a whole. This poses a particular problem to modifications that are supposed to improve the integrated behaviour of an MPCS.

A typical, real-life performance indicator, for example, may encourage a foreman to keep certain machines continually loaded. In many situations, it can be shown that a full utilisation of a machine leads to a degradation of the production lead time of the factory [51,80]. The factory would profit from a schedule that would keep the machines idle at specific points in time [52]. However, it will be difficult to convince the foreman to work according to such a schedule as long as the performance indicators that apply to him encourage him to keep the machines busy all the time.

We feel that further development and implementation of the performance indicators in Appendix B could be a very important contribution to eliminate these sub optimising practices. Appendix B does not necessarily contain the best conceivable performance indicators but is an attempt to describe some performance indicators that apply to the MPCS as a whole as well as a hierarchy of performance indicators for individual

MPCS components that support the performance indicators for the MPCS as a whole.

A final remark regarding the practical application is a trivial but often overlooked one: the basic production processes of a manufacturing organisation should be under control before an attempt is made to introduce or improve automated MPCSs.

15.3   Relation with Other Work

The development of the Reference Model for MPCSs has been influenced considerably by empirical observations of existing MPCSs. However, the Reference Model is not an abstract representation of any existing MPCS. It is a *design* of an idealised, integrated, and flexible MPCS. It is unlikely that an existing MPCS is exactly organised according to the Reference Model, although bits and pieces of it may be in place.

Hierarchical Organisation.   The Reference Model proposes a hierarchical organisation of MPCS controllers. How does this compare with other publications?

Almost all papers published in workshops about manufacturing control [1,97] propose hierarchical control systems. In fact, so many different hierarchies are proposed that it looks as if it is not important what a hierarchy looks like. We disagree.

The vast majority of the proposals fail to describe *tasks* of controllers. They only mention that manufacturing organisations have a hierarchical *structure*. Such proposals may be intuitively appealing but lack any justification.

In some cases, task definitions are given or suggested. However, they are based on criteria that are specific for the configuration and physical implementation of an individual MPCS. Hence, they are not generally applicable. We find, for example, hierarchies based on planning horizons. Control tasks are classified according to the period of time their decisions are valid. This will be different in each MPCS: one MPCS uses forecasts of demand six months ahead of time, the other three. Other proposals seem to be based on the geographical distribution of machines on the shop floor. All machines within a certain area of the shop floor are part of a Workstation! So many Workstations make up a Workcell, etc. It is obvious that such criteria are different everywhere and make no sense in terms of control.

We are not aware of another Reference Model for MPCSs that is as complete as the one described here and that is justified on the basis of design criteria discussed in Section 3.4. We are not aware either of a Reference Model for MPCSs that includes management as described in Part III of this work.

Open Systems Interconnection. There is another well-known, and widely used Reference Model, which addresses a different application area: the Open Systems Interconnection Reference Model for computer networks [64,98,104,113]. There are many differences between the OSI Reference Model and the Reference Model for MPCSs due to the differences in application. The MPCS model is inherently more complex because it deals with many aspects of planning that do not apply to computer networks. Further, the OSI model does not have the clean separation between management and execution as our model. OSI treats management as an extension of the basic Reference Model rather than as an implementation, which together with an Executor replaces the basic Reference Model [43,44,101].

Integrated Manufacturing. It should be noted that the idea of 'integrated manufacturing control' is not new and is promoted by many papers on 'Computer Integrated Manufacturing'. Many of these papers involve computer networks that are used to connect controllers with the intention of providing flexibility because controllers can communicate with each other. They fall short, however, in determining which information is to be passed via these networks.

This reflects the misguided, but nevertheless widespread, view that computers and networks are a panacea for manufacturing problems. We concur with Reference [50] and feel that while the technology for manufacturing –including processes and computer hardware and software– is improving rapidly, a basic understanding of the systems issues remains incomplete.

Our approach is therefore different. It makes no sense to try to automate planning and control tasks as long as we do not know which planning and control tasks we need. Try to rigorously and critically analyse what the relevant planning and control tasks are and what information is required by the human or systems that execute those tasks. Once those tasks have been defined, it can be decided whether humans, systems, or both should execute them. The networking is then provided to enable communication between those systems.

15.4   Further Research

This work could be used as a basis for further research in the following areas:

- Validation of the Reference Model by domain experts.

  The model covers many domains that are traditionally the field of experts in

Industrial Operations Research, Industrial Engineering, Robotics, Sensing, Control Theory, Industrial Design, Mechanical Engineering, etc. Their re-thinking of manufacturing planning and control problems in the context of the entire MPCS, as suggested by the Reference Model, could result in pointers for validating the Reference Model.

- Complete functional specifications.

Only a few complete functional specifications for MPCS components have been published. Even fewer have been published for MPCS Management components. The development of these specifications for a variety of applications is an important research area and will also provide feedback for improvements of the Reference Model.

- Methods for mathematical validation.

The Reference Model is a qualitative description of an integrated system. In addition to qualitative descriptions, we would like to have models for the quantitative analysis of the various coordination tasks of an MPCS and for the analysis and approximation of the performance of an MPCS as a whole.

- Control algorithms for integrated manufacturing planning and control systems.

The Reference Model poses important constraints on the potential algorithms executed by MPCS controllers.

The majority of the control algorithms published are not based on constraints and objectives imposed by a large-scale system as the MPCS. The development of planning and control algorithms 'within the framework' of the Reference Model could result in a suite of coherent algorithms and thus further substantiate the approach towards integrated manufacturing control.

Appendix A

Interaction Primitives

We introduce interaction primitives in section 2.2 and use them throughout this document to illustrate the global task definitions of MPCS and MPCS Management components.

By giving interaction primitives, one can model the interactions of systems only partially. One should also specify when which interactions take place.

The concept of interaction primitives is closely linked with the concept of parallel processes. Parallel processes are agents that perform tasks concurrently. These agents interact to share information to affect each other's behaviour. These interactions can be modeled by interaction primitives. Thus, we model how a Company Controller and an MPCS interact by exchanging commands and status reports. The commands and status reports carry information that affects the activities of the Company Controller and MPCS.

From a mathematical point of view, the information shared by the parallel processes represents coupled constraints or variables, which affect multiple computations, i.e. the computations executed by the interacting, parallel processes. Thus, we model how a decision of a Company Controller that certain products be dispatched affects the decisions of an MPCS. The lists of products to be dispatched are coupled variables shared by both controllers as a result of an exchange of interaction primitives that identify the products.

Our definition and interpretation of interaction primitives is based on the semantics of the interaction concept in the formal specification language LOTOS [17,31,65,107]. The most important characteristics of the interaction primitives are described below.

- The interaction primitives define the exchange of information at a high level of abstraction. They describe the meaning of the information rather than its format or coding in messages. They only describe the information that is exchanged. They do not describe how the information is exchanged, how the information is generated, or how the information is directed.

- Interaction primitives merely describe that two systems establish a set of values to which both can refer. Both systems may have their own constraints on the values that are acceptable to them. If the interaction takes place, values are established that are acceptable to both. The interaction primitives do not describe, however, the individual constraints of the systems.

  Table 2.1 defines, for example, a primitive to define the production target for the MPCS. Both the Company Controller and MPCS may have constraints on the feasible targets. The MPCS, for example, can only produce a limited volume of products. If the interaction, modeled by the interaction primitive, takes place, a target is established that satisfies all constraints. The values of the parameters of the target are shared both by the Company Controller and MPCS. The interaction primitive abstracts from possible negotiations of the Company Controller and MPCS to determine feasible production targets, which in reality might precede the MPCS's commitment to a target.

- The interaction primitives are executed instantaneously. The primitive in Table 2.1 to define a production target can therefore be used to model an MPCS that does not ship products to Customers while it negotiates the production targets with the Company Controller. If it is desirable to model that an MPCS ships to Customers during its negotiations, we would need an interaction primitive to model the Company Controller proposing a production target, one to model the MPCS's response, and one to model the shipping of products between the Company Controller's proposal and MPCS's response.

Appendix B

Performance Guidelines for MPCS Controllers

Performance guidelines for MPCS controllers are introduced in chapter 11. They can be expressed in several forms. One could, for example, define inventory as the number of products, as a commercial value, as the percentage of the planned consumption, etc. One could express the processing lead time per product type, per group of product types, or as a weighed average [87]. In addition to performance guidelines of the MPCS, we could define performance guidelines for MPCS Management, such as the time needed to prepare the MPCS for new production targets [87], or the total production costs [1].

We will discuss some performance guidelines for the logistic MPCS controllers. Each of the performance guidelines can be defined in two ways:

1. How well does the MPCS component execute the commands it receives? For example, how reliably does it meet the due dates, or how reliable is the material consumption?

2. Which commands could the MPCS component potentially execute, in what order, combination, and frequency? For example, how many parts of one type can a Cell/Line Controller process, which combinations of different part types, and how often can it receive new commands?

Table B.1 summarises the performance guidelines. It lists some of them as 'cost guidelines', which express performance in terms of events that should not happen or quantities that should be kept small. negative performance guidelines or cost guidelines. A

---

[1]One could also define guidelines indicators to indicate the degree to which MPCS controllers should really use the Management information. It may occur, for example, that the actual product structure used does not reflect the BOM [87].

brief explanation follows.

The performance of an MPCS Executor can be expressed by:

- Order cycle time. How much in advance should products be ordered by the Company Controller?

- Service level. How many orders have been realised in time?

- Target frequency and lead time. How often can an MPCS Executor receive new targets?

- Mix and volume. Which products and in what quantities can the Factory produce?

The following cost factors depend on the operational control of a Factory Controller:

- Inventory levels. The levels of the inventories controlled by the Factory Controller.

- Cell/Line utilisation. Does the Factory Controller utilise the Cell/Lines' capacities to their fullest potential?

The performance of a Cell/Line can be expressed by:

- Throughput. How many parts can a Cell/Line produce?

- Job lead time. How quickly should a Cell/Line execute a job?

- Lead time reliability. Does the Cell/Line meet its commitments to execute jobs within the specified time slots?

- Order frequency. How often can a Cell/Line be ordered to execute jobs?

- Mix and volume flexibility. How many different items can the Cell/Line produce? Can the Cell/Line produce the required volumes or does it produce batches?

The following cost factors depend on the operational control of a Cell/Line Controller:

- Work-In-Progress. How many and which parts are contained by a Cell/Line to achieve a certain throughput?

Controller	Performance Guidelines	Cost Guidelines
Factory Controller	Order cycle time Service level Mix & volume	Inventory levels Cell/Line utilisation
Cell/Line Controller	Throughput Job lead time Lead time reliability Order frequency Mix & volume flexibility	Work-In-Progress Workstation utilisation
Workstation Controller	Operation lead time Part quality	Automation Module utilisation
Note: The accuracy of status feedback is a performance guideline for each controller.		

Table B.1: Performance And Cost Guidelines For Logistic MPCS Components

- Workstation utilisation. Does the Cell/Line Controller utilise the Workstation's capacities to their fullest potential?

The performance of a Workstation can be expressed by:

- Operation lead time. How much time does a Workstation need to execute an operation? What is the variability?

- Part quality. What is the quality of parts processed by a Workstation?

Appendix C

Impact of Production Parameters on MPCS Performance

Production parameters are introduced in chapter 10. They have a profound impact on the performance of an MPCS. We discuss key characteristics of production parameters that particularly affect the MPCS performance and summarise them in Table C.1.

Topology of Inventory Points.   Production parameters should define the topology of inventory points so that a Factory Controller can meet customer demand within the customer order cycle time, while limiting costs of inventories. The specifics of a given topology have consequences for the costs and performance of an MPCS.

- Commonality of items. Items with a high commonality, in stocks controlled by a Factory Controller, have a demand whose variance is reduced in comparison with the demand of products, provided that the demands for the latter are not correlated. The commonality can therefore be exploited to predict more accurately the demand for items in a stock and hence reduce stock levels.

  Commonality can be disadvantageous in special cases, for example, if the common parts would replace specialised but cheaper parts.

- Value of items. The costs of maintaining stock depends on the inventory levels and the value of the items. It also depends on physical characteristics of the items. Some items require regulated environments. The costs of space depends on the size of the items. Some items may require specialised handling to avoid contamination, etc.

- Job lead times. The time it takes to process items determines how long in advance their demand must be estimated. Short lead times generally allow more reliable

214

estimates of demand and therefore lower inventory levels.

- Variability of job lead times. Typically, it is important that the job lead times have little variability.

- Structure of precedence constraints. It may be required that process plans allow certain precedence constraints, such as alternative paths, parallel, or sequential tasks.

Cell/Line Allocation. We describe below some characteristics of Cell/Line allocation that impact the MPCS performance.

- Independence of distinct Cell/Lines. The execution of a job by one Cell/Line should not affect the lead time of a job executed by another Cell/Line. Jobs that cannot be executed independently should therefore be allocated to the same Cell/Line.

- Utilisation of Cell/Lines. The allocation of jobs to certain Cell/Lines affects their utilisation.

- Specialisation of Cell/Lines. Cell/Lines may be specialised to execute a class of jobs. It is preferable to allocate such classes of jobs to these specialised Cell/Lines.

Bill Of Materials. We describe below some characteristics of a BOM that affect a Cell/Line's ability to process items in the required time slots.

- Invariance of operation lead times. It may be required that a Workstation executes an operation within a specified time window irrespective of the combination and sequence of its operations.

- Value generation by operations. Not all operations add value to a part. Examples of non-value adding operations are tests, transportation, material handling and storage, and tool handling [11]. Non-value adding operations can be eliminated but not without costs. There is therefore a trade-off between investments and potential savings.

- Relative work contents. A Cell/Line Controller may be able to synchronise operations better, and hence reduce the Work-In-Progress, if the operations have equal work contents (measured in time units).

A Cell/Line Controller may be able to achieve low WIP levels and a high utilisation of Workstations despite changes in work contents. However, it should be able to find a suitable sequence for the operations [10].

- Absolute work contents. It may be required that work contents have a certain absolute value. The duration of manual operations, for example, affects the variability in the operation time, the average quality of the operation, and the time needed to learn to perform the operation.

- Commonality of parts. It may be advantageous if different products can be manufactured with common parts. This reduces the number of different parts to be designed, stored, transported, processed, etc. However, it may as well be economical to have specialised parts.

- Reliability of operations. It may be required that the cycle time and material consumption of operations are accurately predictable.

Bill Of Process. The following characteristics of BOPs affect a Cell/Line's performance.

- Self-sufficiency of Cell/Lines. We already mentioned that Cell/Lines should not affect each other's capacity. One can ensure that this is the case by allocating disjunct groups of Workstations to different Cell/Line Controllers. The mutual dependence of Cell/Lines is further reduced if the Workstations of each Cell/Line are self-sufficient in the sense that they exchange much more parts, much more frequently among Workstations associated with the same Cell/Line than with Workstations of other Cell/Lines.

  Cell/Lines with independent capacities can be realised in other ways as well. It would require complex schemes of Cell/Line Controllers sharing the same Workstations.

- Alternative allocations. Distinct Workstations can be suited to execute the same operations. A Cell/Line Controller can then command the Workstation that happens to be available or that does not require a set-up to execute the operation.

- Utilisation of Workstations. The allocation of operations to certain Workstations affects their utilisation.

- Specialisation of Workstations. It may be desirable that Workstations are specialised to execute a class of operations. It may be worthwhile to allocates these operations to specialised Workstations.

Recipe.   The following characteristics of Recipes can affect the MPCS performance.

- Alternative sequence of processing steps. A Workstation Controller may be able to reduce its operation lead time if it can select from alternative process steps. It may select those steps, for example, for which Automation Modules are available.

- Number of test instructions. It may be required that recipes prescribe that a Workstation should execute test instructions to ensure high quality operations or early detection of defective parts.

- Recovery procedures. It may be required that a recipe prescribes how a Workstation should recover from failures to execute an operation. It may, for example, try to place a component again if the previous attempt failed.

- Set-up activities. It may be important that a Workstation spends little time to change over to another operation. The recipe should describe minimum activities to acquire tools and not require process steps for a particular operation.

Automation Module Allocation.

- Utilisation. It may be required to allocate certain process steps to certain Automation Modules to change their utilisation.

- Specialisation. It may be desirable that Automation Modules be specialised in the execution of a few process steps. Classes of process steps should be allocated to specific Automation Modules.

Production Parameter	Key characteristics
Topology of Inventory Points	Commonality of stock-items
	Value of stock-items
	Job lead times
	Job lead time variability
	Structure of precedence relations
Cell/Line Allocation	Independence of capacities of distinct Cell/Lines
	Utilisation of Cell/Lines
	Specialisation of Cell/Lines
Bill Of Materials	Invariance of operation lead times
	Value added by operations
	Relative work contents
	Absolute work contents
	Commonality of parts
	Reliability of operations
	Alternative precedence constraints
	Parallelism in operations
Bill Of Process	Self-sufficiency of Cell/Lines
	Alternative allocations
	Utilisation of Workstations
	Specialisation of Workstations
Recipe	Alternative sequences of processing steps
	Test instructions
	Recovery procedures
	Non-value adding handling/processing
	Amount of set-up activities
Automation Module Allocation	Utilisation of Automation Modules

Table C.1: Major Characteristics Of Production Parameters

Appendix D

Common Misconceptions in Understanding Reference Models

In chapter 3, we define what a Reference Model is and discuss its role. The meaning and role of Reference Models have not yet been widely discussed in literature. For the sake of clarity and a correct interpretation of this work, we give some common interpretations of a Reference Model for MPCSs that do *not* conform to the one used in this work:

- A Reference Model is *not* a collection of computers. The model describes abstract processes that execute certain tasks. It does not prescribe how these processes are physically implemented by humans, machines, computers, etc. It does neither prescribe or preclude that multiple processes can be implemented on the physical system.

- A Reference Model is *not* a picture with boxes and arrows. These pictures only illustrate the model. Its essence is captured by descriptions that explain the tasks and configuration of the controllers.

- The hierarchy described by the Reference Model does *not* reflect a geographical grouping of machines. The different levels in the hierarchy are not the result of such a simple grouping mechanism as:

    - three Workstations form a Cell/Line;
    - six Workstations form two Cell/Lines;
    - three Cell/Lines form an MPCS;
    - etc.

The levels are distinguished because they execute different tasks. A Cell/Line Controller, for example, has the responsibility to schedule Workstations. It may have to schedule three Workstations, or six, or any other number.

- The hierarchical levels described by the Reference Model are *not* defined by means of time horizons. One cannot say, for example, that a Workstation has to execute its operations within a minute, that a Cell/Line has to execute its jobs within a day, or that an MPCS has to realise its production targets within a month. These time horizons will be different for actual MPCSs and are not a basis for generally applicable definitions.

- A component of the Reference Model is *not* a physical machine. For example, one cannot point to a physical machine and tell that it is an implementation of a Workstation unless one knows which tasks the physical machine can execute. The machine is a Device if it can accept commands to change joint variables, an Automation Module if it can accept commands to process objects, a Workstation if it is used to execute operations that are scheduled, etc.

- Different levels in the hierarchy described by the Reference Model do *not* describe an MPCS at different levels of abstraction. One does view an MPCS at different levels of abstraction if one interprets it as a producer of products, as a control system, as a collection of machines and control computers, as atoms and electrons, etc. In these cases, one views the *same* system but has different *interpretations*. The interpretations at high levels of abstraction consider the function of the system; the interpretations at low levels of abstraction consider the physical implementation. All levels in the Reference Model have the same level of abstraction, describing the tasks of MPCS components. They are *distinct* system components that perform separate tasks concurrently and that *exchange* information.

- The hierarchical structure of the Reference Model does *not* represent a calling tree of a computer program. A calling tree outlines which sub routine is called by which sub routine or program to perform calculations while the other sub routines are not being executed. The Reference Model depicts inherently parallel controllers, each executing its own task but interacting with others to collectively realise the function of an MPCS.

- The Reference Model does *not* reflect the hierarchical nature of many control algorithms. These control algorithms solve a problem step by step. A hierarchical scheduling algorithm could, for example, first calculate the allocation of tools to machines, and subsequently the allocation of parts to these machines.

The Reference Model assigns all tasks relating to the scheduling of operations to one component. The model characterises the scheduling task but does not prescribe whether or not scheduling algorithms should be executed in a hierarchical fashion.

- The Reference Model should *not* be a vague definition that should be sharpened and made more specific for each application. The model is intended to be a rigid, accurate, complete, and generally applicable specification at the appropriate level of abstraction. The Reference Model defines, for example, an abstract concept 'operation'. It tells what an operation is, and gives constraints and requirements for operations in real situations. It is not necessary to redefine what an operation is for a new application. However, the model does not define whether a specific operation results in a television or a radio. Such definitions have to be provided when the model is implemented.

Subject Index

Bibliography

[1] *Proc. on Factory Standards Model Conference.* National Bureau of Standards, Washington D.C., 1985.

[2] R. Akella, Y. Choong, and S. Gershwin. Performance of Hierarchical Production Scheduling Policy. *IEEE Trans. on Components, Hybrids, and Manufacturing Technology*, CHMT-7(3), 1984.

[3] J. Albus, A. Barbara, M. Fitzgerald, E. Kent, C. McLean, H. Bloom, L. Haynes, C. Furlani, E. Barkmeyer, M. Mitchell, H. Scott, D. Blooquist, and R. Kilmer. A Control System for an Automated Manufacturing Research Facility. In *Robots 8 Conference*, 1984.

[4] J. Albus, A. Barbara, and R. Nagel. Theory and Practice of Hierarchical Control. In *Proc. Twenty-third IEEE Computer Society International Conference*, 1981.

[5] J. Albus, H. McCain, and R. Lumia. NASA/NBS Standard Reference Model for Telerobot Control System Architecture (NASREM). Technical report, National Bureau of Standards, Technical Note 1235, 1987.

[6] K. Baker, M. Magazine, and H. Nuttle. The Effect of Commonality on Safety Stock in a Simple Inventory Model. *Management Science*, 32(8), 1986.

[7] A. Barbara, M. Fitzgerald, J. Albus, and L. Haynes. RCS, the NBS Real-Time Control System. In *Robots 8 conference*, 1984.

[8] S. Benson, F. Biemans, M. Denver-Dudley, R. Goffin, M. Harden, R. Kruijk, M. Ronayne, S. Sjoerdsma, J. van den Hanenberg, and A. van der Pol. CAM Reference Model. Technical report, Philips, CFT-Rep. 13/87, 1987.

[9] J. Beukeboom, F. Biemans, C. Hehl, S. Sjoerdsma, and H. van Veen. CAM Reference Model. Technical report, Philips, CFT-Rep. 01/89, 1989.

[10] K. Bhaskaran. Mixed Model Scheduling in Flow Lines. Technical report, Philips Laboratories-Briarcliff TN-89-017, 1988.

[11] K. Bhaskaran. Selecting Process Plans. Technical report, Philips Laboratories-Briarcliff TN-89-001, 1988.

[12] F. Biemans. Distributed Internal Transport Control. Technical report, Philips, CFT rep. 01/84, 1984. M.Sc. Thesis, Twente University, The Netherlands.

[13] F. Biemans. The Design of Distributed Transport Systems as a Major Standard Interface in Computer Integrated Manufacturing. *Computers In Industry*, 7(4):319–333, 1986.

[14] F. Biemans. A Reference Model of Production Control Systems. In *Proc. IECON86, IEEE Industrial Electronics Society*, 1986.

[15] F. Biemans. A Trial Architecture of a Workstation Formally Specified. Technical report, Philips CFT-Report 02/86, 1986.

[16] F. Biemans. Management of Manufacturing Planning and Control Systems. Technical report, Philips Laboratories-Briarcliff, TR-89-008, 1989.

[17] F. Biemans and P. Blonk. On the Formal Specification and Verification of Computer Integrated Manufacturing Architectures using LOTOS. *Computers In Industry*, 7(6):491–504, 1986. Also published as Philips Centre for Manufacturing Report 32/86.

[18] F. Biemans, T. Harosia, F. Holmquist, C. Lee, A. v.d. Stadt, and R. Wendorf. EAASY-0 Specifications. Technical report, Philips Laboratories-Briarcliff, Tech. Rep. TR-88-048, 1988.

[19] F. Biemans and S. Sjoerdsma. The Toyota Production Organisation. Technical report, Philips, TEO Note 128e, 1985.

[20] F. Biemans and S. Sjoerdsma. Description and Motivation of a Workstation Controller Architecture, towards integration and standardisation. Technical report, Philips CFT-Rep 55/86, 1986.

[21] F. Biemans and S. Sjoerdsma. Functional Specifications for a Cell/Line Controller–to be published. Technical report, Philips Laboratories-Briarcliff, 1989.

[22] F. Biemans and C. Vissers. Computational Tasks in Robotics and Factory Automation. In *Proc. of 1988 IEEE Workshop on Special Computer Architectures for Robotics and Automation of the IEEE Robotics and Automation Conference*, pages 1–27, 1988. reprinted in Computers in Industry, Vol 10., No.2, 1988, pp. 95-112, also to appear in Sistemi E Automazione.

[23] F. Biemans and C. Vissers. A Reference Model for Manufacturing Planning and Control Systems. *Journal of Manufacturing Systems, Vol 8, No 1*, 1989.

[24] F. Biemans and C. Vissers. A Unified Systems Perspective on the Management and Execution of Manufacturing Planning and Control. *to be published*, 1989.

[25] H. Bijl and A. v. Strien. CN-Station Controller Project, product outline. Technical report, Philips, I&E Automation Systems, 1988.

[26] G. Blaauw and F. Brooks. Computer Architecture, 1982. lecture notes, Twente University of Technology, The Netherlands, and University of North Carolina at Chapel Hill.

[27] P. Blonk and F. Biemans. An Architecture of Internal Transport Control Systems - towards standardisation of the functional behaviour of internal transport systems, appendices. Technical report, Philips CFT-Rep. 10/87, 1986.

[28] P. Blonk and F. Biemans. An Architecture of Internal Transport Control Systems - towards standardisation of the functional behaviour of internal transport systems. Technical report, Philips CFT-Rep. 09/87, 1987.

[29] L. Boza, C. Chavous, J. Hsu, R. Lake, and M. Pratt. CIMA-An Integrated Architecture for Product Realization. *ATT Technical Journal*, pages 17–25, jul-aug 1986.

[30] H. Bremermann. Complexity and Transcomputability. In R. Duncan, editor, *The Encyclopedia of Ingnorance*. Pergamom Press, 1978.

[31] H. Brinksma. A Tutorial on LOTOS. In M. Diaz, editor, *Proc. of. the IFIP WG 6.1 5th. Int. Workshop on Protocol Specification, Testing and Verification*. North-Holland, 1985. also published as: Provisional LOTOS tutorial ISO/TC 97/SC 21 N.

[32] H. Brinksma, G.Scollo, and C.Vissers. Experience with and future of LOTOS as a Specification Language. In *Proc. of. the Third CCITT SDL Forum, The Hague*. North-Holland, 1987.

[33] F. Brooks. *The Mythical Man-Month*. Addsion-Wesley, Reading, Mass., New York, 1975.

[34] R. Brooks. A Robust Layered Control System for a Mobile Robot. *IEEE Journal of Robotics and Automation*, RA-2(1):14–23, 1986.

[35] A. Clark and H. Scarf. Optimal Policies for a Multi-Echelon Inventory Problem. *Management Science*, pages 475–490, 1960.

[36] G. Cohen. Main Components of a Computer-Integrated Manufacturing Reference Model. *CIM Review*, pages 28–36, 1988.

[37] Computer Integrated Manufacture-Open System Architecture/AMICE Consortium. CIM-OSA: A Primer on Key Concepts and Purpose, 1985. 489 Avenue Louise, B 14-B-1050, Brussels, Belgium.

[38] P. Courtois. On Time and Space Decomposition of Complex Structures. *CACM*, 28(6):590–603, 1985.

[39] M. Curley, M. Moor, W. van Schaik, P. Trouwen, and A. Wiersma. CAM Architecture for BU Display Components. Technical report, Philips, proprietary information, 1988.

[40] G. Domeingts. Decisions in Production Control. In *Proc. CAM-I Factory Management Workshop, Aachen, West Germany*, 1980.

[41] L. Dorst and P. Verbeek. The Constrained Distance Transformation: A Pseudo-euclidean, Recursive Implementation of the Lee-Algorithm. In *Signal Processing III: Theory and Applications*. Elsevier Science Publishers (North-Holland), 1986.

[42] R. Beukeboom et al. CAM-SKE 90, CAM en Sectorbesturing, philips, proprietary information, 1989.

[43] K. Farrington, B. Potter, and A. Pras. PANGLOSS/12 – Network Management Requirements of a Gateway. Technical report, ESPRIT, 1987.

[44] International Organization for Standardization. Information Processing Systems – Open Systems Interconnection - Basic Reference Model - part 4: Management Framework, DIS 7498-4. 1988.

[45] B. Fox and K. Kempf. Complexity, Uncertainty, and Opportunistic Scheduling. In *Proc. IEEE Conf. AI. Appl.*, pages 487–492, 1985.

[46] S. French. *Sequencing and Scheduling, an Introduction to the Mathematics of the Job-Shop*. Ellis Horwood, Ltd., 1982.

[47] J. Galbraith. *Designing Complex Organizations*. Addison-Wesley Publishing Company, 1973.

[48] G.Boot, J.Bosmans, R.Dermaut, A.Duijf, P. van Houtte, M. de Jong, and H.Lammers. Proposal for a C.I.M. Information Architecture for Consumer Electronics. Technical report, Philips, 1987.

[49] Y. Gerchal, M. Magazine, and B. Gamble. Component Commonality with Service Level Requirements. *Management Science*, 34(6):753–760, 1988.

[50] S. Gershwin, R. Hildebrant, R. Suri, and S. Mitter. A Control Theorist's Perspective On Recent Trends in Manufacturing Systems. In *Proc. 23rd Conference on Decision and Control*, 1984.

[51] E. Goldratt. Cost Accounting: the Number One Enemy of Productivity. In *APICS 26th. Annual International Conference Proceedings*, pages 433–435, 1983.

[52] E. Goldratt and R. Fox. *The Race*. North River Press, 1986.

[53] S. Graves. A Review of Production Scheduling. *Operations Research*, 29(4), 1981.

[54] S. Graves. Safety Stocks in Manufacturing Systems. *Journ. Manuf. Oper. Mgt*, 1:67–101, 1988.

[55] M. Groover. *Automation, Production Systems, and Computer-Aided Manufacturing*. Prentice-Hall, Inc, 1980.

[56] J Habers. Definitions Equipment Monitoring Consumer Electronics. Technical report, Philips Consumer Electronics, 1988.

[57] I. Ham. Group Technology Applications for Higher Manufacturing Technology, Introduction. Tech. Note Pennsylvania State University.

[58] A. Hanson and E. Riseman. VISIONS: A Computer System for Interpreting Scenes. In A.Hanson and E.Riseman, editors, *Computer Vision Systems*. Academic Press, 1978.

[59] A. Hax and D. Candea. *Production and Inventory Management*. Prentice-Hall, 1984.

[60] S. Hoekstra and J. Romme. *To Integral Logistic Structuring*. Kluwer, 1985. (in Dutch).

[61] T. Hopp and K. Lau. A Hierarchical Model-Based Control System for Inspection. In Gardener, editor, *Automated manufacturing ASTM STP 862 L.B.*, pages 169–187. American Society for Testing and Materials, 1985.

[62] ISO/TC184/SC5/WG1. The Ottawa Report on Reference Models, version 0.1. Technical report, 1986.

[63] ISO/TC184/SC5/WG1. Reference Model for Shop Floor Production Standards, Part 1: A Reference model for Standardization and a Methodology for Identification of Standards Requirements. Technical report, 1989.

[64] ISO/TC97/SC16. Information Processing Systems, Open Systems Interconnection, Basic Reference Model, International Standard iso/is 7498. Technical report, ISO, 1983.

[65] ISO/TC97/SC21. A Formal Description Technique Based on the Temporal Ordering of Observational Behaviour, Draft International Standard iso/dis8807. Technical report, ISO, Information Processing Systems, Open Systems Interconnection, 1985.

[66] A. Jones and C. McLean. A Production Control Module for the ARMF. National Bureau of Standards, Washington, DC 20234.

[67] R. Karp. Combinatorics, Complexity, and Randomness. *Communications of the ACM*, 29:98–110, 1986.

[68] E. Kent and J. Albus. Servoed World Models as Interfaces Between Robot Control Systems and Sensory Data. *ROBOTICA*, 2:17–25, 1984.

[69] S. Kochbar and R. Morris. Heuristic Methods for Flexible Flow Line Scheduling. *Journal of Manufacturing Systems*, 6(4):299–314, 1988.

[70] A. Kusiak and G.Finke. Selecting Process Plans in Automated Manufacturing Systems. *IEEE Journal of Robotics and Automation*, 4(4):397–402, 1988.

[71] L. Langenhoff. An Exact Decomposition Method for Multi-Echelon Inventory Systems. Technical report, Technical University Eindhoven-M.Sc. Thesis (in Dutch), 1988.

[72] L. Langenhoff and H. Zijm. An Analytical Theory of Multi-Echelon Production/Distribution Systems. *Statistica Neerlandica*, 1989.

[73] C. Lin and C. Moodie. An Energy-Saving Production Scheduling Strategy for Hierarchical Control of Steel Manufacture. Technical report, Purdue Laboratory for Applied Industrial Control, report nr. 147, 1985.

[74] O. Maimon. Real-Time Operational Control of Flexible Manufacturing Systems. *Journal of Manufacturing Systems*, 6(2):125–136, 1987.

[75] J. Markvoort, M. de Pont, F. van der Heijden, and N. Kemper. Fast and Accurate Manipulator Modules, Instruction Set. Technical report, Philips Centre For manufacturing Technology Report, 27/87, 1987.

[76] M. Mesarovic, R. Erlandson, D. Macko, and D. Fleming. Satisfaction Principle in Modeling Biological Functions. *Kybernetes*, 2:67–75, 1973.

[77] M. Mesarovic, D. Macko, and Y. Takahara. *Theory of Hierarchical Multilevel, Systems*. Academic Press Inc., 1970.

[78] R. Milner. *A Calculus of Communicating Systems*. Springer-Verlag, 1980. Lecture Notes in Computer Science.

[79] Y. Monden. *Toyota Production System*. Institute of Industrial Engineers, 1983.

[80] J. Nijenhuis and A. de Kok. Simulation Study of BIC-Production at Valvo RHW, Hamburg. Technical report, Philips, Centre for Quantitative Methods, proprietary information, 1988.

[81] Danish Standards Organization. The Framework for Development of a Generic Reference Model for CIM Systems, 1986. Contribution of the Danish Member Body to ISO/TC184/SC5, Document S142/CIM(DK).

[82] S. Panwalker, R. Dudek, and M. Smith. Sequencing Research and the Industrial Scheduling Problem. In *Symposium on the Theory of Scheduling and its Applications, Lecture Notes in Economics and mathematical systems, Springer Verlag, NY*, 1973.

[83] R. Parker and R. Rardin. Complexity Theory: Concepts, Results and Implications for Discrete Optimization. Technical report, Lecture Notes, Columbia University, Industrial and Systems Engineering, Rep. J-81-9, 1981.

[84] D. Parnas. On the Criteria To Be Used in Decomposing Systems into Modules. *CACM*, 15(12), 1972.

[85] H. Van Dyke Parunak and J. White. A Synthesis of Factory Reference Models. In *Proc. IEEE Workshop on Languages for Automation, Vienna, Austria, August 1987*, 1978.

[86] H. Pels and J. Wortman. Decomposition of Information Systems for Production Management. In *Proc. of IFIP W.G. 5.7 Working Conference on Decentralized Production Management Systems*, 1985. also published in 'Computers In Industry' Vol.6, Nr. 6, pp. 435-453.

[87] Factory Systems Consumer Goods project. The Four Phase Approach, part 6, Logistics Performance Measuring Programme. Technical report, PHILIPS, 1986.

[88] F. Rodammer and K. White. A Recent Survey of Production Scheduling. *IEEE trans. on systems, man, and cybernetics*, 18(6), 1988.

[89] P. Sanders. Parts Feeding In Flexible Automation. Technical report, Philips Centre For Manufacturing Technology, CTB 83.55.004, 1983.

[90] P. Scharbach. Formal Methods and the Specification and Design of Computer Integrated Manufacturing Systems. In *Proc. Int. Conf. on The Development of Flexible Automation Systems*. The Institution of Electrical Engineers, 1984. Conference Publication Number 327.

[91] R. Schonberger. *Japanese Manufacturing Techniques, nine hidden lessons in simplicity*. The Free Press, New York, 1982.

[92] SECS-I and SECS-II. Semi Equipment Communications Standard, data link and transaction (proposed), 1985.

[93] S. Seward, S. Taylor, and S. Bodlander. Progress in Integrating and Optimizing Production Plans and Schedules. *Int. J. Prod. Res*, 23(3):609–624, 1985.

[94] S. Shingo. *Study of Toyota Production System, from Industrial Engineering Viewpoint*. Japan Management Organization, 3-1-22 Shiba Park, Minato-Ku, Tokyo, 1981.

[95] E. Silver and R. Petersen. *Decision Systems for Inventory Management and Production Planning*. John Wiley & Sons, 1985.

[96] H. Simon. *The Sciences of the Artificial*. The MIT Press, Cambridge, Mass., 1981.

[97] *Cell Controllers*. Society for Manufacturing Engineers and its Computer and Automated Systems Association, April 26-27, 1988.

[98] A. Tanenbaum. *Computer Networks*. Prentice-Hall Inc., 1981.

[99] J. Thistle and W. Wonham. Control Problems in a Temporal Logic Framework. *Int. J. Control*, 44(4):943–976, 1986.

[100] P. van Eijk, C. Vissers, and M. Diaz (eds.). *The Formal Specification Language LOTOS, Results of the ESPRIT/SEDOS Project*. North-Holland, 1989.

[101] C. Vissers. Basic (architecture) concepts of the OSI Reference Model. In *Proc. Telematica-3*. SIC-Nederlands Genootschap Informatica, 1986. Dutch.

[102] C. Vissers and L. Logrippo. The Importance of the Service Concept in the Design of Data Communication Protocols. In M. Diaz, editor, *Proc. of. the IFIP WG 6.1 Int. Workshop on Protocol Specification, Testing and Verification*, pages 3–17, 1986.

[103] C. Vissers, G. Scollo, and M. van Sinderen. Architecture and Specification Style in Formal Descriptions of Distributed Systems. In *Proc. of. the 8th. IFIP WG 6.1 Int. Workshop on Protocol Specification, Testing and Verification*. North-Holland, 1988.

[104] J. Voelcker. Helping Computers Communicate. *IEEE Spectrum*, pages 61–70, March, 1986.

[105] T. Vollman, W. Berry, and D. Whybark. *Manufacturing Planning and Control Systems*. Richard D. Irwin, Inc, 1984.

[106] R. Wendorf. Implementation Notes on the First Generation Briarcliff Workstation Controller. Technical report, Philips Laboratories, TN-88-121, 1988.

[107] R. Wendorf. Implementation of Manufacturing Planning and Control Systems from LOTOS Specifications. Technical report, Philips Laboratories, TN-88-015, 1988.

[108] J. Wijngaard and J. Wortmann. MRP and inventories. *European Journal of Operational Research*, 20:281–293, 1985.

[109] R. Wittrock. An Adaptable Scheduling Algorithm for Flexible Flow Lines. *Operations Research*, 36(3), 1988.

[110] R. Yeomans, A. Choudry, and P. ten Hagen. *Design Rules for a CIM System.* North-Holland, 1985.

[111] Y. Zhang and R. Paul. Robot Manipulator Control and Computational Costs. In *Proc. of the 1988 IEEE Workshop on Special Computer Architectures for Robotics and Automation*, 1988.

[112] H. Zijm and A. de Kok. Production Planning and Inventory Management in a Telecommunication Industry. In *A. Chikan (ed), Proc. 4th. Int. Symposium on Inventories.* Elsevier, 1988.

[113] H. Zimmermann. OSI Reference Model-The ISO Model of Architectures for Open Systems Interconnection. *IEEE Trans. Commun.*, COM-28, 1980.

Summary

Goal. We substantiate a systematic and methodical approach to manufacturing planning and control. The result is a Reference Model for 'Manufacturing Planning and Control Systems' (MPCSs), which allows us to develop integrated and flexible MPCSs. We define an MPCS as a system that can accept production targets, and manufactures products accordingly, provided that it is supplied with the resources needed to manufacture these products.

The systematic approach to MPCSs differs from approaches that address planning and control problems in isolation. These latter approaches do not define or validate the control problems in terms of the relevance of the objectives and constraints in the context of the overall system. These approaches may therefore not lead to performance and flexibility improvements of an MPCS as a whole.

It is difficult to develop an 'integrated MPCS' because an MPCS is concerned with many problems that are poorly understood in terms of their interrelationships. Inventory control, product design, scheduling, process planning, robot control, transport control, layout design, and product inspection, for example, are related to each other in several ways. Hence, the effect of a specific activity on the performance and flexibility of the entire MPCS can hardly be assessed.

In addition to the requirements that an MPCS be an integrated system, we require that an MPCS be flexible. This implies that the MPCS can easily be prepared to operate in new applications, with different product portfolios, production volumes, or production costs.

Approach. We discuss the development of an integrated and flexible MPCS. As a prelude to the actual development, we select a development strategy.

We first have to describe an MPCS as a black box, from the perspective of a commercial department that outlines production targets and budgets to the MPCS.

This systems view is important because it specifies the external functionality of the system we intend to develop.

The inherent complexity of an MPCS forces us to begin its development at a high level of abstraction, where we consider its functionality rather than its physical implementation. Further, we have to decompose the black box MPCS into MPCS components that can be understood with more ease.

We adopt a two-pronged technique to decompose the MPCS. On the one hand, we analyse the interactions of the MPCS and its environment. This serves to identify the tasks executed by the MPCS to process and generate the information exchanged in the interactions. On the other hand, we apply 'separation of concerns'. This means that relatively independent subtasks of an MPCS are assigned to separate MPCS components. This leads to relatively independent components that can be considered as distinct entities in the context of the entire MPCS and therefore can be understood more easily.

The top-down analysis, which starts from the black box MPCS and derives a structure of interacting MPCS components with the above described decomposition techniques, can lead to integrated and flexible MPCSs.

Before attempting to design a 'flexible MPCS', which can honour requests to change its product portfolio, production volumes, or production costs, we design an 'inflexible MPCS', which has fixed product portfolio, production volume, and production costs.

Inflexible MPCS.   We apply the decomposition strategy step by step. In each step, we identify MPCS components which have to be decomposed further in a subsequent step. Each decomposition requires an analysis of such problems as:

- the control of inventory levels to be able to quickly dispatch products while keeping inventory costs low;

- the scheduling of operations;

- the coordination of machines to execute operations;

- the determination of the trajectories of joints of machines;

- the servoing of joints; and

- related feed back tasks.

We analyse which information is required to solve the problems, which information can be supplied by other components, which information can be neglected, which problems are relatively independent, etc. This analysis and the application of the 'separation of concerns' maxim guide the decomposition process. We also describe real-life scenarios and show how the MPCS components can play a role in these scenarios.

The result of the decomposition is the 'Reference Model for MPCSs'. A Reference Model for MPCSs describes an organisation of functional components of an MPCS, independent of and without reference to any physical means with which it may be implemented. This organisation has two aspects: *structure* and global *tasks*. A structure identifies components that co-operate by exchanging information to affect mutual behaviour. A global task characterises a component's essential mission, its domain of responsibility.

A Reference Model for MPCSs serves two purposes. First, to obtain an abstract specification of the MPCS as a basis for the detailed design of the functional components. Second, to provide those who wish to discuss MPCSs with a reference for a common terminology and concepts.

Flexible MPCS. It appears that the tasks of the components of an inflexible MPCS, as defined by the Reference Model, are constrained by some parameters. They depend on the product portfolio, production capacity, or production costs, or, in other words, on the 'application' of the MPCS. The application changes slowly in comparison with the production targets. The parameters can therefore be viewed as given constraints during the time a components of an inflexible MPCS execute orders. However, they may change over a longer period of time in a flexible MPCS.

We therefore replace the black box description of an inflexible MPCS with a black box description of a flexible MPCS. Subsequently, we implement the flexible MPCS by a system that executes the control tasks, the 'MPCS Executor', and by a system that generates the application-specific parameters, 'MPCS Management'. MPCS Management can provide the MPCS Executor with the application-specific parameters and thus prepare it to operate in changed applications. When loaded with the information specific for a particular application, the MPCS Executor can execute the tasks of an inflexible MPCS in that particular application.

The MPCS Executor and MPCS Management are inherently complex systems. We have to consider them at a high level of abstraction and have to decompose them to reveal their internal organisation.

It appears that the MPCS Executor can be decomposed rather easily: its internal structure reflects the structure of the inflexible MPCS.

To decompose MPCS Management, we analyse the interactions of MPCS Management with the MPCS Executor and the commercial department and apply the 'separation of concerns' technique. The decomposition requires analysis of such problems as:

- feasibility analysis of production targets;

- product design;

- machine design;

- process planning;

- development of control procedures;

- maintenance;

- monitoring, etc.

The result of the decomposition is a Reference Model of MPCS Management.

Application.   Both, the Reference Models of MPCS and MPCS Management, are indispensable mile stones in developing an integrated MPCS. They define components of the total, complex MPCS, and meanwhile allow us to keep sight of the relationships between the separate components and between the components and the MPCS as a whole.

The Reference Models are also indispensable milestones on the way towards flexible MPCSs. The effected decompositions are based on the analysis of a wide variety of practical applications and are therefore valid for these applications. Moreover, MPCS Management, which serves to realise general applicability, has been analysed in considerable detail.

The Reference Models can be used as:

- blue prints to compare existing MPCSs with the idealised MPCS defined by the models;

- top-level specification for the development of systems that execute planning and control tasks; and

- a definition of constraints and objectives for planning and control algorithms.

The development of planning and control algorithms 'within the framework of the Reference Model' could result in a coherent suite of planning and control algorithms. These will further contribute towards integrated manufacturing control.

The most important contribution of this study is the application of design heuristics that ensures an MPCS that has the conceptual integrity, clarity, and modularity, which allow a human to master and understand an MPCS as a whole despite its formidable, inherent complexity. The advantage, and uniqueness, of the ability to understand an MPCS should not be underestimated. It allows us to quickly develop MPCSs, to control them, to adapt them, to monitor them, etc. These are qualitative but nevertheless important characteristics of an MPCS in addition to more traditional, quantitative performance criteria.